TECHNICAL JUSTICE

Scott Tilley, Ph.D.

Technical Justice

Cover design © Scott Tilley
Cover artwork © Alexander Limbach /Shuttersock
Back artwork © Phonlamai Photo /Shutterstock

Published by CTS Press

CTS
Press

An imprint of Precious Publishing, LLC

Precious Publishing
www.PreciousPublishing.biz

ISBN-13: 978-1-951750-00-8
ISBN-13: 978-1-951750-01-5 (ebook)

TABLE OF CONTENTS

DEDICATION

To all my faithful readers of Technology Today.

PREFACE

On a personal level, 2018 was a year of significant transition. I retired as an emeritus professor from the Florida Institute of Technology. My father passed away. After 8-plus years, I stopped writing my weekly "Technology Today" column for FLORIDA TODAY. In many ways, it was a year of hard endings, but as the old saying goes, when one door closes, another door opens.

From a technology perspective, 2018 was another turbulent year. It's challenging to pick a handful of particularly noteworthy developments, but if pressed, I'd select three that really stood out: big data, privacy, and security; genetic engineering; and artificial intelligence (AI). Each topic touches on the theme of this book: technical justice.

In our digital lives, we are throwing off data like a dog throws off water droplets while shaking when getting out of the rain. We don't know how much data there is, it flies everywhere, and we don't seem to care about it. The resulting increase in big data means that the number of data breaches seems to be increasing daily. Privacy is dead, and security a mirage. If you are hacked, where can you go to obtain redress? If someone installs ransomware on your computer, what can you do? What forms of justice are available to the average person?

We now have babies being engineered at the genetic level, using tools such as CRISPR. There is little societal oversight to this wild-west of biomedical engineering. Our legal system has not caught up with technical advances in the field. Who owns your DNA data? After you die, can you be simulated in a digital afterlife? There are now concerts featuring holograms of dead stars. Who is concerned with the original person's best interests? Where is the justice in entertainment companies making money off of their reputations?

In the past, most jobs lost to automation were blue-collar manufacturing positions. Now, AI is starting to take over more white-collar jobs, and I believe this phenomenon has just begun. As AI improves much faster than we can adapt, the impacts on society will be profound. Most professionals don't belong to a union, so where will they go when their job is taken over by a robot? Ironically, AI has the legal profession clearly in its sights. Case law can be ingested by machine learning algorithms much faster than interns can perform background research. Ironically, lawyers may find there is little justice available to them when they are replaced by an algorithm. Would we be comfortable with an AI-powered judicial system?

Change is rarely easy, but it is always inevitable. In the past year, I've moved on to new projects and am enjoying my new-found freedom. But rest assured, I'm not abandoning the technology field. There are several books already in the pipeline. If you want to be kept informed of my current and future activities, join the mailing list by going to www.cts.today/join and filling out the simple form.

I always welcome your feedback. I can be reached via email at TechnologyToday@srtilley.com, on Twitter @TechTodayColumn, and Facebook: www.facebook.com/stilley.writer. You can also learn more about all of my writing at www.amazon.com/author/stilley.

Scott Tilley
Melbourne, FL
October 21, 2019

ACKNOWLEDGMENTS

This is Volume 8 in the "Technology Today" series, the final collection of my newspaper columns written for FLORIDA TODAY. I've been incredibly fortunate to receive so much positive correspondence from my loyal readers since I started the column way back in October 2010. Without their support, I would not have had such an enjoyable time. Thank you!

Thanks to Dave Dickinson for his suggestion of the book's title. I polled members of writing groups, and his idea really resonated with me. Technical justice is elusive today; tomorrow, who knows.

As always, thanks to my family and my growing brood of pets for reminding me every day that there's more to life than work.

Calendar for Year 2018 (United States)

January
S	M	T	W	T	F	S
	1	2	3	4	5	6
7	8	9	10	11	12	13
14	15	16	17	18	19	20
21	22	23	24	25	26	27
28	29	30	31			

○:1 ◐:8 ●:16 ◑:24 ○:31

February
S	M	T	W	T	F	S
				1	2	3
4	5	6	7	8	9	10
11	12	13	14	15	16	17
18	19	20	21	22	23	24
25	26	27	28			

◐:7 ●:15 ◑:23

March
S	M	T	W	T	F	S
				1	2	3
4	5	6	7	8	9	10
11	12	13	14	15	16	17
18	19	20	21	22	23	24
25	26	27	28	29	30	31

○:1 ◐:9 ●:17 ◑:24 ○:31

April
S	M	T	W	T	F	S
1	2	3	4	5	6	7
8	9	10	11	12	13	14
15	16	17	18	19	20	21
22	23	24	25	26	27	28
29	30					

◐:8 ●:15 ◑:22 ○:29

May
S	M	T	W	T	F	S
		1	2	3	4	5
6	7	8	9	10	11	12
13	14	15	16	17	18	19
20	21	22	23	24	25	26
27	28	29	30	31		

◐:7 ●:15 ◑:21 ○:29

June
S	M	T	W	T	F	S
					1	2
3	4	5	6	7	8	9
10	11	12	13	14	15	16
17	18	19	20	21	22	23
24	25	26	27	28	29	30

◐:6 ●:13 ◑:20 ○:28

July
S	M	T	W	T	F	S
1	2	3	4	5	6	7
8	9	10	11	12	13	14
15	16	17	18	19	20	21
22	23	24	25	26	27	28
29	30	31				

◐:6 ●:12 ◑:19 ○:27

August
S	M	T	W	T	F	S
			1	2	3	4
5	6	7	8	9	10	11
12	13	14	15	16	17	18
19	20	21	22	23	24	25
26	27	28	29	30	31	

◐:4 ●:11 ◑:18 ○:26

September
S	M	T	W	T	F	S
						1
2	3	4	5	6	7	8
9	10	11	12	13	14	15
16	17	18	19	20	21	22
23	24	25	26	27	28	29
30						

◐:2 ●:9 ◑:16 ○:24

October
S	M	T	W	T	F	S
	1	2	3	4	5	6
7	8	9	10	11	12	13
14	15	16	17	18	19	20
21	22	23	24	25	26	27
28	29	30	31			

◐:2 ●:8 ◑:16 ○:24 ◐:31

November
S	M	T	W	T	F	S
				1	2	3
4	5	6	7	8	9	10
11	12	13	14	15	16	17
18	19	20	21	22	23	24
25	26	27	28	29	30	

●:7 ◑:15 ○:23 ◐:29

December
S	M	T	W	T	F	S
						1
2	3	4	5	6	7	8
9	10	11	12	13	14	15
16	17	18	19	20	21	22
23	24	25	26	27	28	29
30	31					

●:7 ◑:15 ○:22 ◐:29

Jan 1	New Year's Day	May 28	Memorial Day	Nov 12	Veterans Day (observed)
Jan 15	Martin Luther King Jr. Day	Jun 17	Father's Day	Nov 22	Thanksgiving Day
Feb 14	Valentine's Day	Jul 4	Independence Day	Dec 24	Christmas Eve
Feb 19	Presidents' Day	Sep 3	Labor Day	Dec 25	Christmas Day
Apr 1	Easter Sunday	Oct 8	Columbus Day (Most regions)	Dec 31	New Year's Eve
Apr 13	Thomas Jefferson's Birthday	Oct 31	Halloween		
May 13	Mother's Day	Nov 11	Veterans Day		

LOOKING AHEAD TO 2018

AR, automated assistants, and blockchain

January 5, 2018

I'd do a much better job at predicting the future if I had the proper technology to do so. Alas, there's no app for that yet, so instead, I'm left gazing into the pundit's crystal ball to see what exciting new developments might occur in 2018. There are plenty of possibilities, but three in particular might be ready for prime time: augmented reality, automated assistants, and blockchain.

Augmented Reality: The December 8 issue of "Entertainment Weekly" featured Dwayne "The Rock" Johnson on the cover, dressed in a Santa suit, holding a large candy cane, and appeared to be singing. If you used the free "Life VR" app, the cover sprang to life: The Rock was indeed singing, speaking and moving around. There were several other examples of embedded video content in the magazine. Although the app has "VR" in its name, it's not virtual reality (VR): it provides an augmented reality (AR) experience, where video and animation are superimposed on real-world imagery and seen using your smartphone or special glasses.

Apple is one company that is investing heavily in AR technologies. They even provide a sophisticated developer's kit as part of the latest iOS release. I expect to see a flurry of AR apps coming this year. Florida-based Magic Leap is another company to watch in this space.

Automated Assistants: I can't remember a time when I didn't have Alexa at home. I use Amazon.com's disembodied assistant in the Echo on a daily basis. She's not perfect, but she does get more capable all the time. I've come to expect her presence so much that when I'm in an environment without her, such as my car, I find traditional user interfaces cumbersome. Why should I take my eyes off the road to fiddle with knobs or poke at a wonky touchscreen just to change the channel while driving, when Alexa could do it for me so much better?

I expect that automated assistants like Alexa will continue to infiltrate our lives in 2018. They can be embedded in just about every appliance and device we use. Soon we'll wonder how we ever lived without them – which is not necessarily a good thing.

Blockchain: The technology that makes bitcoin and other forms of cryptocurrency possible is called blockchain. It's a distributed ledger system that provides a transparent and secure record of global transactions. Many companies and government agencies are looking into blockchain for purposes other than digital money. For example, IBM has proposed that the Canadian province of British Columbia use blockchain to manage the supply chain in the legal marijuana market. Quite timely, given recent legislation in California.

# # #

WORLD ECONOMIC FORUM

A global meeting of thought leaders in Davos

January 12, 2018

Each year, the World Economic Forum holds a high-level meeting in Davos, Switzerland with some of the biggest names in the technology field. We're talking royalty, presidents and prime ministers, and CEOs of the world's largest corporations. My invitation to Davos seems to have been lost in the mail, so instead of flying to Europe for a live report, I'll comment on some of the key topics likely to be discussed at the Forum in 2018. These topics are related to cybersecurity, artificial intelligence, and biotechnology.

A Digital Geneva Convention: The Geneva Convention was created to limit atrocities during wartime. After the chemical attacks of World War I, the global community felt that there were some things outside the bounds of even the most strident conflicts. Most countries have signed on to the Geneva Convention, although there have been numerous instances of the convection having been violated.

There is now a movement to create a digital version of the Geneva Convention that would limit the use of cyber-weapons. For example, targeting civilian infrastructure with massive attacks like the WannaCry malware that affected numerous facilities last year would be out of bounds under a digital Geneva Convention. It's a laudable goal, but cyberattacks are a very different type of warfare than conventional clashes of nation states.

Managing AI's Risks and Rewards: There is no doubt that AI has the potential to positively affect many areas of our economy. It's also true that AI could seriously disrupt large segments of our society. A superintelligence that might emerge from our own research into intelligent machines could easily become ungovernable – to the detriment of the lesser beings. That is to say, us.

There is movement towards crafting a code of ethics for AI systems, which would impose human moral boundaries on thinking machines. The question is, which ethical framework to choose? The world is replete with very different civil structures.

Intelligent and Connected Bio-Labs: The rapid developments in the underlying technologies supporting genetic engineering and other forms of biotechnology are quickly creating the necessary conditions for engineering bioweapons of unprecedented destructive power. The dissemination of information related to these intelligent bioweapons has been aided by the growth of the Internet and the dark web.

For example, there are only two official storage facilities that still maintain live strains of the smallpox virus: one in the United States, and one in Russia. However, the genetic code for smallpox is accessible to anyone online. With this knowledge, recreating live smallpox would be a relatively simple matter of manufacturing.

The global elite at Davos should have plenty to talk about. Solutions would be nice too.

# # #

BIG DATA AND PRIVACY

We throw off data like a dog shakes off water

January 19, 2018

We are connected to many devices and services all the time. The connections can be wireless, such as Bluetooth, cellular, or Wi-Fi. They can also be online or through a smartphone app, such as when using Facebook or other social media programs.

These connections are increasing in number all the time, driven in part by the Internet-of-Things (IoT). The result is that we are throwing off data like a dog throws off water droplets while shaking when getting out of the rain. We don't know how much data there is, it flies everywhere, and we don't seem to care about it. But we should. The data we voluntarily provide while enjoying our online lives provides an incredibly accurate digital fingerprint of all our activities – including what we do in the real world. This has serious privacy implications.

Very few people are aware of just how much can be deduced from your online activities. A quick Google search will reveal everything from your current and past addresses to your employment history. Private investigators pay a small amount to uncover much more personal information that is gathered by data brokers and sold online. What companies do with your private data is outlined (albeit obliquely) in their terms of service, but these agreements are often pages long and written in legalize that is designed to be hard to understand for the average user.

It is already the case that government agencies in China are using aggregations of personal data to create a social credit score, rather like a financial credit score. Your financial credit is used to determine eligibility for things like loans and rental opportunities. The social score is much broader, and can be used in predictive analytics to determine eligibility for social services, job openings, and generally track your activities.

There is another source of personal data that we provide, albeit unwillingly. Government programs such as Section 702 of the Foreign Surveillance Intelligence Act give U.S. intelligence agencies the power to spy on electronic communication and phone calls of foreigners residing outside the USA. This is common practice for many countries. But if a U.S. citizen is traveling abroad, there's a very good chance that their communications will be scooped up in this vast digital net as well. In essence, the privacy of U.S. citizens and legal residents become collateral damage in the war on terror.

This is a classic example of the tradeoffs between the benefits of big data collection and analysis, and the trampling of personal privacy in our connected society. The irony is that we willingly participate in giving away our data much of the time.

# # #

BIG DATA AND SECURITY

Privacy is dead and security a mirage

January 26, 2018

The massive Equifax data breach was a top technology story of 2017. According to a Federal Trade Commission notice released last September, "If you have a credit report, there's a good chance that you're one of the 143 million American consumers whose personal information was exposed. The breach lasted from mid-May through July 2017. The hackers accessed people's names, Social Security numbers, birth dates, addresses and, in some instances, driver's license numbers, credit card numbers, and dispute documents with personal identifying information."

How was such a stunning data breach possible? The answer is depressingly simple: Equifax did not to patch their website, even though they were previously warned of known vulnerabilities. In other words, human error that the hackers exploited.

Are such lapses common? Again, depressingly, yes.

In December 2017, CNET reported another data breach affecting 123 million US households. The leak occurred in October on an unsecured cloud-based server that was left online by marketing analytics company Alteryx. Ironically, the repository contained data belonging to Experian, an Equifax competitor.

It's no better in 2018. Just yesterday, Bell Canada announced a data breach affecting 100,000 customers.

Complexity is the bane of security, and today's big data systems are exceedingly complex. A major difference between hacks a few years ago and today's attacks is that the data is extensive and stored online, not on private networks. Anyone with the right tools or skills can get access to the treasure trove – and they do it all the time.

These data breaches are primarily software-based, but hardware is not immune. Two devastating microprocessor security vulnerabilities, called "Spectre" and "Meltdown," were recently disclosed by Google researchers. Chips from Intel, AMD, and ARM are potentially affected by this design exploit, going back a decade or more. These chips are in everything we use: smartphones, laptops, tablets, and so on. The exploit allows the capture of sensitive data on the chip, including passwords and cryptographic keys.

I have little doubt the vulnerability has been exploited by national security agencies worldwide for some time. If I was truly paranoid, I'd say the design flaw was devilishly deliberate. A cunning plan to hide a back door for those with the keys to gain unfettered access.

In the era of big data today, privacy is dead and security a mirage.

# # #

PERSON OF INTEREST

The state of surveillance

February 2, 2018

Our personal data is used by companies and government agencies for a variety of purposes. Sometimes we benefit from their services, such as the recommender systems used by Amazon.com and Netflix that provide genuinely useful insights. Other times, data brokers aggregate the information about our online lives and sell it to everyone from malicious hackers to marketing firms.

The organizations acquire this data from us in three ways: we volunteer it (e.g., social media), it is captured by sensors from the devices we use (e.g., smartphones), or it is stolen from us (e.g., websites and services we use are hacked). Irrespective of the manner in which the data is captured, the result is that we are entering a new era of surveillance.

The TV show "Person of Interest" used a fictional computer, called "The Machine," and its evil twin, called "Samaritan," to illustrate the potential uses and abuses of an artificial super intelligence that could track everything that everyone did at all times. The computer was fed data from many sources and used predictive algorithms to determine possible future outcomes for individuals and the state. The citizens of New York City went about their daily lives, blissfully unaware that they were being watched.

Today, the real world has caught up with fictional television. The state of surveillance has advanced in technical capabilities in recent years, so much so that we are indeed being watched almost all the time. And for the most part, we're OK with it, because we're not aware it's happening. But it is, and in some places, more than you thought possible.

The city of Guiyang in southwest China has a population of about 4.5 million. It is a laboratory for the government's plan to build the world's largest and most sophisticated camera surveillance system. Every resident of Guiyang (as with most of China) is required to have an ID card. Their image is stored in a central database so that they can be found and tracked by the millions of cameras installed throughout the area.

Each camera has enough artificial intelligence that the system can determine your identity, even when driving a car, and can accurately estimate your age, your gender, your ethnicity, and your height and weight. It can even identify you based on your gait. A BBC correspond gave the system a test by providing his photograph and then walking around town. The police surrounded him in seven minutes.

# # #

FALCON HEAVY

An epic launch

February 9, 2018

Epic. Awesome. Incredible. Exciting. Inspiring.

These are just some of the adjectives that spring to mind from the recent SpaceX launch of the Falcon Heavy. I watched liftoff via FLORIDA TODAY online, but once the rocket was accelerating away from the pad, I ran outside to see bright orange flames streaking upwards. Not since the final launch of the shuttle in 2011 do I remember feeling so thrilled at witnessing a daytime blastoff.

As amazing as the launch was, the nearly simultaneous landing of the two side cores was even more unbelievable. I could hear the multiple sonic booms coming from Cape Canaveral all the way at Florida Tech. Watching the two rockets falling from the sky and then settling into perfect landings made the hair on the back of my neck stand up. It truly looked like we were being invaded from Mars.

I showed video clips of the Falcon Heavy to my students, even though the classes had very little to do with space travel. It was just too good of a teachable moment to pass up. What better way to inspire aspiring engineers than to witness what's possible when you have a great team of innovators led by a master persuader?

I also showed the students photos of the rocket during assembly and test, to get a better feel for its size. 230 feet is really tall when you see little people standing nearby. The business end of the rocket, bristling with 27 Merlin engines, is a sight to behold. Only the archival footage we watched of the giant Saturn V rocket that went to the moon, powered by five massive Rocketdyne F-1 engines, can compare in scale and grandeur.

The central core was supposed to land on the drone ship a few hundred miles out in the Atlantic Ocean. It seems the rocket missed its target by about three hundred feet and slammed into the water at 300mph. According to Elon Musk, preliminary analysis suggests only one of three engines fired to slow the rocket down before landing. But even this minor miss can't put a blemish on an otherwise stupendous achievement.

How could the launch and return possibly get any better?

By sending a Tesla Roadster with a Starman dummy at the wheel into deep space of course!

Images of the Roadster in orbit around Earth have gone viral. It's an unreal scenario, yet it's very moving. Even though the car overshot it's intended Mars orbit and is now heading out towards the asteroid belts, it's still a marvelous example of our engineering talents and creative aspirations.

There are problems in the world, but there is greatness too.

# # #

IMAGE

Scott Tilley finds NASA's lost satellite

February 16, 2018

I've been quite busy lately, but it seems I've been even busier than I thought. A few weeks ago, while indulging in some amateur astronomy, I was searching for the lost ZUMA mission. This secret payload was launched on a SpaceX Falcon 9 from Cape Canaveral on January 8, 2018. The launch was successful, but the satellite appeared to vanish. Given its classified mission status, no one was really sure what happened.

I never did find out. But what I did find instead was NASA's long-lost IMAGE satellite. IMAGE was launched in 2000 from Vandenberg Air Force Base and went dark in 2005. Its mission was to study changes in Earth's magnetosphere caused by solar wind. By 2007, NASA had failed to reestablish communication with the satellite and declared the mission over.

I spoke to representatives at NASA's Goddard Space Flight Center in Maryland, and they confirmed that the signals I detected were from IMAGE. Hopefully, they will be able to reconnect with the satellite and revive the system enough to continue its mission. The rebirth of a dormant satellite doesn't happen very often, so I'm pleased that I was able to be a part of it.

When the news broke of my discovery, I was contacted by numerous media outlets. Most wanted to know how, exactly, I had found the IMAGE satellite. A few wanted to know more about my background. Only one reporter asked if I was the right person with whom to speak.

There's a huge danger is making assumptions based on unverified information. In the big data world, we know that predictive analytics will produce erroneous results if the input training data is incorrect.

The reporters knew my name, Scott Tilley. A quick Google search for me finds numerous links to my academic and professional work. The search also shows that my Ph.D. is from the University of Victoria, which is in British Columbia, Canada. The news reports said the amateur astronomer was from British Columbia, so clearly, I must be the guy.

Except it wasn't me.

I'm no amateur astronomer. I don't even own a telescope, never mind the radio hardware and software that modern astronomers use to scan the skies.

I did use to live in British Columbia, but that was over 20 years ago.

I do work in technology, but I'm a professor, not an "electrical technologist."

I was very tempted to play along and say it was indeed me who found IMAGE. But that wouldn't be fair to my 47-year-old namesake in Roberts Creek, British Columbia. He should enjoy his 15 minutes of fame.

Scott, give me a call sometime. We should talk.

# # #

CODED COUTURE

A great example of STEAM

February 23, 2018

Until recently, I had not thought about the influence of technology on clothing and fashion. However, the arts are often where innovative use of technology occurs first, and the current exhibit at the Ruth Funk Center for Textile Arts at Florida Institute of Technology exemplifies this interplay.

The exhibit is called "Coded Couture." It is an interactive experience that "explores the idea of customized fashion in the digital age." I was privileged to have a private showing this week, led by Keidra Navaroli, Assistant Director of the Funk Center. The Executive Director and Chief Curator of University Museums at FIT, Carla Funk, also accompanied us on the tour.

There were several pieces that I found fascinating. There is an augmented reality display that greets you as you enter the exhibit space. There is a feathered headpiece that moves as you get closer, almost as if it was reacting to your presence. There is a dress with a metal frame that pulses with different patterns based on your speech patterns – an interesting lie detector in real time. Two pieces of clothing move and react to where a camera is pointing at them – which is a stand-in for responding to someone's gaze and where they are staring at you. There is also a skirt that can display tweets sent to it in real-time.

When your clothes have their own Twitter account, you know technology has disrupted the fashion world.

One particularly thought-provoking piece is by Amy Congdon. She blends traditional techniques like embroidery with bio-technology such as tissue engineering and 3D bio-printers as part of the Biological Atelier project, which is set in the year 2080. According to Congdon's website, "The project envisions a world where materials are not made, they are grown. What if we could manipulate our body to grow seasonal jewelry? Cosmetic surgery is replaced by tissue-engineered disposable grafted skin embellished with precious stones."

Each piece in the exhibit is categorized according to one of four different types of coding: biological, cultural, psychological, or synergistic. Actual programming does not have such a delineation, but it's an interesting way of thinking about computing's place in the real world.

This is Engineer Week, and I can't think of a better way for engineers to expand their understanding of what it means to be an engineer than to pay a visit to the "Coded Couture" exhibit. It's a great example of STEAM (STEM + Arts) that offers a different perspective on the role of technology in society. It might even provide opportunities for truly novel senior design projects through cross-disciplinary collaboration.

# # #

ROBOT COMEDY

Is this thing on?

March 2, 2018

Over the holidays, I had a few standup comedy gigs. Stop laughing — that's not the punchline.

I do comedy at senior centers around town and up in Canada to bring a little cheer to people who may not have many visitors, or who may not be able to get out much. For them, the holidays are not a happy time. When I started doing comedy several years ago, I knew I was going far outside my comfort zone. But I figured if I could turn a few frowns upside down, it would be worth it. After all, the worst that could happen would be that people wouldn't laugh and I'd make a fool of myself. I do that in class all the time, so no worries there.

My comedy style is Seinfeld-esque: I'm primarily a storyteller, with (what I hope are) witty observations of daily life. It seems to go over quite well. The last performance I had last year was just after Christmas, and over 60 seniors laughed out loud for nearly an hour.

So, what does comedy have to do with technology?

I've written in the past about the possible dangers posed by advanced robots powered by artificial intelligence taking jobs away from people across our society. But until recently, I never thought about robot comedy. Could a robot be a standup comedian?

Doing comedy is not easy. The writing is difficult, the delivery is challenging, and you have to adjust your demeanor in real-time in response to the audience's reactions. But there is a certain playbook you follow for the whole process, which means it's theoretically possible to codify the playbook and have it enacted by a machine.

Indeed, it's already been done. Researchers at Carnegie Mellon University have a robot stand-up comedian named Data that has been featured in TED talks.

Unbeknownst to me, there's an entire area of artificial intelligence research dedicated to comedy. If we want to prepare robots to interact with humans in the near future, as equals and not servants or overlords, the robots will have to adopt some of our social mannerisms, and humor plays an essential role in inter-personal communication. But it's actually quite difficult to describe how humor works.

Can a robot write funny material for a comedy routine? Yes. That also has been done, and some of the material is quite good, if a bit quirky. Companies like Narrative Science already sell news-writing software that can be the basis for different forms of writing, including comedy.

Here's one from the UK-based Joking Computer: "What do you get when you cross a frog with a road? A main toad."

Is this thing on?

# # #

Swarm Intelligence

Autonomous micro-drones are coming your way

March 9, 2018

When you think of drones, what images pop into your head?

Do you think of military drones, like the Predator, that are deployed in distant places? These are large drones that are remotely controlled by pilots on the other side of the world. They are effective weapons against insurgents and hard-to-reach targets.

Do you think of hobby drones, like the ones children use to fly around the house? These toys are controlled by a radio console or a smartphone app. They have limited range, but they are inexpensive and fun to use.

Many other types of drones are somewhere between toys and Predators in terms of size, cost, and capabilities. But one thing that almost all drones share is that they are single machines flying under human control. They are effectively airplanes that have removed the pilot to a safe distance.

The next generation of drones may remove the pilots altogether.

Some of the most interesting developments in drone research is focused on the creation of many small drones that fly autonomously (without a pilot) but in concert with one another. They act somewhat like a swarm, not individuals. In fact, they mimic many of the characteristics of biological swarms, such as a flock of birds or a school of fish. There is no single individual in charge, but as a collective, the swarm manages to achieve its goals.

If you've ever witnessed Canada geese during migration, you know they fly in a V formation. There is a definite structure to their skein. On the other hand, a murder of crows seems to have little structure; each crow flies in what looks like a rather haphazard manner, but they do make progress. The lesson here is that there is more than one way for a swarm to operate.

The U.S. Department of Defense has already done experiments with autonomous micro-drones (just 6.5 inches long) called Perdix. These drones are meant to be used for unmanned aerial surveillance. The drones have been dropped in canisters from fighter jets in canisters, which then open automatically to deploy the drones.

The drones dynamically adjusting their configuration for optimal performance, even if several drones are lost. Arguably, the drone swarm demonstrates a level of artificial intelligence previously unseen in such complex environments. The basis for their intelligence is called bio-inspired computing.

How would you react to a swarm of insect-sized drones buzzing your way?

# # #

SLAUGHTERBOTS

When AI meets drones, remarkable things can happen

March 16, 2018

The recent passing of Stephen Hawking has led to considerable coverage of his incredible scientific contributions – and rightly so. His work on black holes and quantum gravity put him in the same league as Einstein and Newton in the pantheon of great minds. His research is all the more remarkable considering the physical challenges he faced while dealing with amyotrophic lateral sclerosis (ALS) for so many years.

I read his book, "A Brief History of Time," shortly after it was first released in 1988. I've always tried to follow advancements in theoretical physics, and his writing made a complex topic very accessible. The book has sold over 10 million copies in 20 years, so clearly his message resonated with people around the world.

More recently, Professor Hawking became concerned about the rapid developments in artificial intelligence (AI). He was an early signatory to 2015 document titled "Research Priorities for Robust and Beneficial Artificial Intelligence: An Open Letter," which was signed by many well-known people who shared his concerns, including Elon Musk and Steve Wozniak. They were advocating for more research into the societal aspects of AI. I share their viewpoints that AI has the potential for dramatic improvements in our lives, but it also has the potential to destroy us. Hawking even went so far as to characterize AI as more of a worry than nuclear weapons.

As an example of what is already possible, last year a video was released showing Stuart Russell, a professor at UC Berkeley, demonstrating what he called "slaughterbots." These are tiny drones that fit in the palm of your hand but have autonomous capabilities due to their embedded AI. He released the micro-drone by tossing it into the air, whereupon it circled back and struck the head of a human-form dummy on stage. The drone penetrated the skull and detonated: it was carrying a small but deadly explosive charge. You can view the video yourself at http://bit.ly/2iLHsXG.

Russell said that swarms of such tiny drones "could kill half a city and are virtually unstoppable." They can operate alone or in groups, and they are able to evade most countermeasures. And unlike nuclear weapons, the infrastructure is left standing and uncontaminated.

The drone was programmed to attack the dummy, but what if it decided to attack another target? The AI in the drone may soon be sophisticated enough for it to change its mission mid-flight. Wither humanity?

# # #

FACEBOOK DATA BREACH

Nothing online is ever truly safe

March 23, 2018

On March 13, 2014, the CEO of Facebook, Mark Zuckerberg, posted a note on his timeline that read, "As the world becomes more complex and governments everywhere struggle, trust in the internet is more important today than ever. ... I've called President Obama to express my frustration over the damage the government is creating for all of our future." Zuckerberg was referring to news that the NSA was spying on U.S. citizens. Note the inclusion of the word "trust" in his post, which he felt the government had violated. He also wrote, "That's why at Facebook we spend a lot of our energy making our services and the whole internet safer and more secure."

Fast forward four years. Yesterday, Zuckerberg posted another lengthy note, this time in response to the data breach involving Facebook and Cambridge Analytica. Zuckerberg wrote, "We have a responsibility to protect your data, and if we can't then we don't deserve to serve you. ... This was a breach of trust between Kogan, Cambridge Analytica and Facebook. But it was also a breach of trust between Facebook and the people who share their data with us and expect us to protect it. We need to fix that."

Indeed.

In the four years that separate the two posts, Facebook has grown incredibly powerful. It's still important for us to ensure governance over federal agencies, but it's now arguably more important for us to enjoy some semblance of trust that the social media and Internet companies we use every day will safeguard our data. Right now, that trust is not there, and the fiasco with Facebook and Cambridge Analytica is just the latest in a long series of data breaches that have exposed personal information that was supposed to be protected.

In this particular case, user data was legally provided to Aleksandr Kogan, an academic researcher at Cambridge University, who then passed on the data to Cambridge Analytica illegally. According to CNET, "Kogan reportedly created an app called 'thisisyourdigitallife' that ostensibly offered personality predictions to users while calling itself a research tool for psychologists. As part of the login process, it asked for access to users' Facebook profiles, locations, ..., and importantly, their friends' data as well." The result was a data mining treasure trove of over 50 million user accounts.

It is true that people didn't have to install the app. They also didn't have to agree to share their Facebook data. But most people just click "OK" as fast as they can to get to the application running.

All I can say is that when online, caveat emptor – because no one else is looking after your best interests other than yourself.

# # #

FACEBOOK USER DATA

How to download everything Facebook knows about you

March 30, 2018

Have you ever wondered precisely what Facebook knows about you? At least, what it's willing to admit that it knows about you? If so, there's a way you can download your Facebook user data and see for yourself.

I did it, and it's quite an eye-opener.

To download your user data, login to Facebook. Go to Settings / General Account Settings, then select "Download a copy of your Facebook data" near the bottom of the screen. A new window will open, asking you to "Download Archive." You will be asked for your Facebook password; then a new window will appear with a message indicating your data archive is being created. You should receive two email messages from Facebook: the first confirming your request, and the second a few minutes later telling you your data archive is ready for download.

When I went through the process, it took just two minutes to generate my data archive. Not bad, considering there are over two billion Facebook users worldwide.

The archive is downloaded as a simple folder, with an "index.htm" file that you open with your browser. When you click on the file, a new window opens with your Facebook Profile information. You can then click on any of the following data categories: Contact Info, Timeline, Photos, Videos, Friends,

Messages, Events, Security, Ads, and Applications.

For example, clicking on "Messages" opens a window with a list of every person you've ever sent a direct message to, and clicking on their name shows you the messages in great detail: when each part of the conversation was sent and all the replies. I repeat: these are all messages you've ever sent since you joined Facebook. Every word.

If you click on "Photos," you'll find every photo you ever uploaded to Facebook. Even the ones you didn't share. You'll also find a link for "Facial Recognition Data," which contains the computed hash code that Facebook uses when automatically tagging you in photos and videos.

The largest category of data is "Timeline," which contains every post you've ever made to Facebook, and every comment on your posts made by other Facebook users. I'm not a prodigious Facebook user, but this was still a very long list. My first "Timeline" post was actually from a friend welcoming me to Facebook; it was made on Sunday, October 26, 2014 at 10:23pm EDT.

Clicking on "Security" will show any changes you made to your security settings. It will also detail every single Facebook login you've ever made: the day and time, your IP address, the browser version and computing platform, and the Cookie stored.

Proof positive that nothing you do online ever goes away.

# # #

Digital Afterlife

What can be done with your data after you're gone?

April 6, 2018

During Easter and Passover, many people take the opportunity to contemplate freedom from bondage and new beginnings – including deep thoughts about death, resurrection, and the afterlife. I won't presume to be an expert on such existential matters, but I am interested in a new chapter in this eternal debate: a philosophical examination of the nature of data and humanity.

With all the news about Facebook leaks and data breaches, people are understandably concerned about the privacy and security of their personal information. In fact, it's somewhat of an open question of who owns your data today. You provide it (willingly or otherwise), social media companies monetize it, and hackers steal it.

But what happens when you're gone? Who owns your data then?

The policies concerning account information and online data of the dearly departed are not something most people think about – until they have to. If someone in your family passes away, do you have the ability to access their accounts without their passwords? Can you download their data, postings, and media? Can you discontinue their online presence, or are they doomed to exist as a virtual ghost forever? These are issues with which the courts are already struggling.

Looking a few years into the future, could your data history be used to create a digital avatar of your past life? How much storage and computing power would be needed to simulate you with enough fidelity to pass the Turing test? How would society react to a perpetual digital afterlife?

I think it would be very tempting for some people to agree to even a poor simulation of a loved one if they could carry on simple conversations with them. This is not that far-fetched: think of a new version of Alexa or Siri, imbued with artificial intelligence, able to download personas and interact much as the original person did. Would you have the strength of will to say no?

What if your consciousness was brought back, with nearly complete data memories, in a body genetically engineered to look and act the way you used to - at any age desired. Would your family welcome you home? What if many years had passed between your passing and revival? What if you were like Dune's Duncan Idaho, a clone purposely remade for as long as others wanted, with no concern for your wishes - how could you ever escape?

###

THE LONG GOODBYE

Providing permanence for something that is slipping away

April 13, 2018

I attended "The Long Goodbye" event at FLORIDA TODAY on Thursday evening. It was an incredibly moving experience. Listening to the speakers share their stories about losing their loved ones to Alzheimer's and other dementias made me appreciative of my current life – and apprehensive of what may lay in the future.

There is a concerted medical effort to address the causes of this debilitating disease, but I fear it will be some time before we see a noticeable improvement in treatment options. Our bodies are living longer, but our brain doesn't seem ready for such longevity.

While medical research continues its quest for a cure, what can today's technology do to help ease the process of the long goodbye?

As with many things, it's instructive to look to our past for inspiration.

I remember as a child watching old 8mm film of my parents when they were just married, and our early family vacations to the ocean. For some reason, taking home movies fell out of favor for a long time, until smartphones became prevalent, and now everyone has a powerful camera and video recorder in their hand all the time. Why not make use of it?

NPR has been archiving stories of people's lives as part of their StoryCorps initiative. These are recordings made by amateurs who interview influential people in their lives. For example, a son talking with his father about some aspect of their early struggles upon immigrating to America.

The StoryCorps model can be used to provide a multimedia record of your family member who is struggling with dementia. If there is any positive aspect to "The Long Goodbye," it's the "long" part: you have some time to prepare for the inevitable, and during this period you can plan for a series of conversations that capture memories. Technology can provide permanence for something that is slipping away.

Recording these conversations can open lines of communication that may have been long shuttered. Most people know very little about the early lives of their parents. This is an opportunity to learn about their formative years, how they fell in love, planning for a new family, buying their first house, their experience at work, their social lives, and so on. Getting to know your parent as a person is a unique opportunity, something that people think they will do "sometime in the future," but if the parent passes away suddenly, the chance is gone.

If you are on "The Long Goodbye" journey with an ailing parent, take advantage of the time to record your story. Put it on a DVD, copy it to a thumb drive, or upload it to YouTube. It will stay with you forever.

# # #

Women in Technology

The keys to success

April 20, 2018

Last week, I had the honor of moderating a panel on "Women in Technology: The Keys to Success." The panel was held as part of the Space Coast Tech Council's quarterly dinner at the Holiday Inn Viera. The Tech Council is part of the Melbourne Regional Chamber.

We had two excellent speakers at the panel session: Tauhida Parveen and Susie Glasgow. Tauhida Parveen is Lead Instructor at Thinkful, a NYC-based startup focused on online education for tomorrow's developers. Previously, she served as University Department Chair of Software Engineering at Keiser University. She holds a Ph.D. in Computer Science from the Florida Institute of Technology and an MBA from the University of Central Florida.

Susie Glasgow is the founder, president, and Chief Executive Officer of Kegman Inc., a certified economically-disadvantaged woman-owned small business and a veteran-owned small business providing scientific, engineering, and technical support services to the US federal government. She is also a founding member and president of the Space Coast Chapter of Women In Defense. Susie retired from the United States Air Force Reserves after 29 years of active and reserve duty.

Both panelists described what they view as the keys to success for their own careers in technology-related (and traditionally male-dominated) fields. They have quite different backgrounds and experience, but many of the challenges they faced were similar.

Tauhida shared three of her lessons learned from working in academia and industry: be self-reliant, don't think of yourself as a woman first but as a person, and take charge of your success.

The panel also discussed how hiring more women in technology is part of an institution's own corporate keys to success. If half the population is not actively recruited (and retained), companies are ignoring a huge source of talent at their peril. Susie spoke about the importance of flexibility in the workplace, something that younger companies seem to be addressing with vigor. It is clear that institutions of all sizes, including government agencies, need to focus on flexibility to ensure their continued success.

# # #

Dad

Modern medical technology and the end of life

April 27, 2018

My dad passed away this week. He went through his own "Long Goodbye," suffering from degenerative Parkinson's disease for many years, rheumatoid arthritis, a massive heart attack and triple bypass, COPD, and finally dementia. He entered palliative care last Thursday and was gone by Monday evening.

The staff at the full-care facility where he lived the last few years of his life were terrific. They did everything they could to make his time there as enjoyable as possible. But watching his slow decline, from an active retiree to a passive patient, was difficult.

Modern medical technology has enabled us to keep a body alive far longer than even a few decades ago, but it hasn't really been able to provide the quality of life that sharply defines who we are. There are important end-of-life discussions that must be held before your loved one slips into their memories; don't keep putting them off.

The sheer number of medications my father took on a daily basis was stunning. As his mental state worsened, he often refused to take his pills, spitting them out like a child. It was both amusing and heart-wrenching at the same time. At what point does the medicine stop being beneficial?

The last technology my dad used was Facetime. It was wonderful to see him smile one last time. Physically, a great distance separated us, but emotionally, we were together again. Video chats are now taken for granted, but they are genuinely transformative technologies that were a long time coming.

I gave my dad a Chromebook when he first entered an assisted living facility over five years ago. He learned to use a Windows PC before that, but having me do remote debugging of his computer became increasingly problematic. I wrote about his experiences with a Chromebook in this column, back in July 2013. The fact that the Chromebook updated itself and didn't require constant security updates made it the perfect computer for him. Sadly, he gradually lost interest in even playing his online card games. The last email I received from him was in early 2014.

I'm now dealing with his digital afterlife. Closing down his email accounts, posting final messages on his Facebook page, and generally ending his online presence. I'm glad I had him change his passwords several years ago so that I could prepare for this very event.

In the end, the best care my dad could get came from my mom holding his hand, whispering in his ear that his family loved him and would be fine when he was gone. No modern drugs or advanced technology, just old-school human contact.

# # #

GENETIC ENGINEERING

Our food, our bodies, our selves

May 4, 2018

Last week I presented "Genetic Engineering" at the final "Tech Talk" of our 2017-2018 season. We had a full house in the Harris Community Auditorium (Foosaner Art Museum) discussing the role genetic engineering will have on our food, our bodies, and our selves (society in general).

When people think of genetic engineering, they often think of comic strips or horror movies that highlight everything that could result from science gone awry. Dilbert has several examples of characters with arms growing out of their heads. The new movie "Rampage" with Dwayne Johnson features three animals grown to incredible size as the result of rogue experiments using CRISPR gene editing technology. And of course, we all know how things turned out in "Jurassic Park."

In some ways, genetic engineering is just nature made faster. For example, we've been creating new types of plants for years through selective crossbreeding. The difference is that nature takes many generations of the plant for the changes to become apparent, even when we guide the process. With genetic engineering, we can alter plant DNA with immediate effect.

It's understandable that people have concerns over such genetically modified organisms (GMO). Most scientists say GMO food is safe to eat. Personally, I start becoming uneasy when the DNA from one species is introduced into the genes of another species. This process is called transgenesis, and it's not fictional. For example, in 2011, Chinese scientists generated dairy cows genetically engineered with genes from humans to produce milk the same as human breast milk. This truly is "The Island of Dr. Moreau" made real.

Genetic engineering promises remarkable advances in modern medicine. Targeted gene therapy is already being used on a limited basis to address particularly challenging conditions, including some forms of cancer. Indeed, without recombinant DNA technology, I would not have the synthetic insulin I need to survive. I'd be relying on cows and pigs.

I can imagine genetic engineering being used to farm human organs and replacement parts, guaranteed not to be rejected by your body, and available at your local "Body Bits R Us" big-box store. If I could buy a new pancreas there, I probably would.

Elon Musk formed the company Neuralink to develop brain-machine interfaces to connect humans and computers. He's concerned that without augmenting our natural abilities, we'll be overwhelmed by artificial intelligence and robots in the near future. Genetic engineering has a role to play here as well, but it's starting down the dangerous road of designer babies and eugenics.

It does make you wonder what the end game of genetic engineering for humanity is. Are we heading towards a society of replicants?

# # #

AI Summit

Artificial intelligence goes mainstream

May 11, 2018

Artificial intelligence (AI) is enjoying a remarkable resurgence. Almost every Fortune 500 company is investing in AI technologies. Venture capital is flowing into startups developing the next generation of AI services. It seems every product has an element of AI embedded in it, from digital personal assistants to movie recommender systems. Even the major players are doubling-down on AI: "Google Research" was recently rebranded as "Google AI."

We've been looking into artificial intelligence for over 50 years. It's an area of research that always seemed to be just a few years away from fruition. Each generation makes bold predictions, and until recently, each generation was wrong. Several things have changed in the last decade that have finally made AI deliver on some of its promises.

The first change is the vastly superior computing power that is now available to anyone doing AI research and development. Cloud computing and special-purpose graphics processing units (GPU) have made massive calculations possible that previously would have been prohibitively expensive.

The second change is a different way of implementing AI algorithms. Instead of relying on rule-based systems that attempt to codify the operational environment, most AI systems today rely on machine learning and large data sets to create adaptive models. These models have proven far more effective at tasks such as computer vision and image recognition.

When I was an undergraduate student, I took a course on pattern recognition. Our main project was to develop a program able to identify hand-written letters. I was never able to get my program to work properly. This was due to the type of recognition algorithm I implemented. I tried to break down each letter into individual pixels on a grid, and then identify edges and letter fragments. I could never implement enough rules to account for human handwriting idiosyncrasies. If I were doing the same project today, I'd use a neural network with a large training data set. The system would learn what each letter looked like without me being explicit about the rules.

This week, the Whitehouse is hosting an "AI Summit" with many of the country's leading companies participating in the event. The underlying reason behind the summit appears to be fear that China will outpace America within the next decade in AI research. China is throwing huge amounts of cash at Chinese AI companies. It reminds me of the alarms that went off in the 1980s when Japan started its "Fifth Generation Computing" project that focused heavily on parallel computing to support advances in artificial intelligence. Some good work did come out of the project, but in the end, it was a few generations too soon.

# # #

GOOGLE DUPLEX

Taking robocalls to a whole new level

May 18, 2018

The Federal Trade Commission's Consumer Information website defines a robocall as, "If you answer the phone and hear a recorded message instead of a live person, it's a robocall." You can have your number placed on "Do Not Call" lists, but not all companies respect your wishes. In fact, the FTC reports that there's been an increase in robocalls because technology has made it easier for scammers to use off-shore services that are beyond the legal reach of the US federal government.

Just last week, USA TODAY reported that the Federal Communications Commission had "approved its largest fine ever – $120 million – against Adrian Abramovich, a Miami man who was found to have placed 96.8 million fraudulent robocalls for vacation deals." The FCC alleges that Abramovich used "neighborhood spoofing" technology that generated fake caller ID data so that the call looks like it's coming from a local number.

Would a robocall still be a robocall if the message is not recorded, but live?

At Google's annual developer conference last week, CEO Sundar Pichai demonstrated a new product called Google Duplex. It's an impressive artificial intelligence (AI) program built using Google's TensorFlow Extended technology to conduct natural conversations over the telephone to perform specific tasks. For now, Google is targeting Duplex to enhance the capabilities of their Assistant program.

The demonstration included Duplex making a hair salon appointment and making restaurant reservations. The conversations were complex, but Duplex handled them perfectly, all by itself. You can watch the video online at https://bit.ly/2KSlH1I.

In each conversation, the person was completely unaware that they were speaking with a program. In that sense, Duplex passed the Turing test, which is itself an amazing feat. In the future, should we legislate that AI programs must identify themselves as such so that we're not fooled into thinking we are interacting with an actual person?

What if Duplex is hacked by telemarketers? It could take scamming to a whole new level. I'm honestly not sure if Duplex would be an improvement or more of an annoyance when compared to today's complex call trees we all suffer from when trying to reach a customer service representative.

As for the FTC's definition of robocall, Duplex is not "live," but it's not a recording either. It's an AI. So, is it a robocall?

Personally, I'd like to have Duplex screen my calls. There's already a bot called Lenny that was designed to fool telemarketers into thinking they've reached a real person. It's very amusing to listen to. With Duplex, when a telemarketer comes calling, they can chat with the AI for as long as they like. Technology can work both ways.

# # #

DREAMS

Do androids dream of electric sheep?

May 25, 2018

Why do we dream?

Dreams can be as vivid as reality, making us question what is true and what is imaginary. Edgar Allen Poe wrote, "All that we see or seem is but a dream within a dream." What secrets do dreams reveal about ourselves? How do we interpret our dreams? Do we dream about events that will inevitably occur, or are they just shades of things that may come to pass?

In 1899, Sigmund Freud published his seminal book, *The Interpretation of Dreams*, which explored some of the common themes we all seem to experience in dreams. Does dreaming about death imply someone you know is about to pass? Why are dreams about falling so common? Why don't we remember more about our dreams when we wake? Is this a coping mechanism for our psyche, or is it a failing of our biological brains?

The truth is, no one knows the answers to these fundamental questions. We've been studying dreams for a long time, but much of the psychological research on dreams remains just theory. Prominent scientists have competing explanations for the role of dreams in our mental health.

These theories remain just that, theory, because we lack the ability to monitor what we dream, or to observe with high fidelity what other people are dreaming. However, that may be about to change.

Imagine there was a way to DVR your dreams each night. In the morning, you could replay a video of your dreams, stopping and rewinding at will. I doubt the recording would match what little you remember, which might hurt or help the therapeutic nature attributed to dreaming.

Now imagine there was a way for someone else to monitor your dreams in real time. Perhaps they could even interact with you while you were dreaming. Would they see images in color? In full depth? Would there be audio? How might they change the path of your dreams? Could this technique be used to cure children of nightmares? Or to reduce the anxiety experienced during recurring dreams by people with PTSD?

This is all hypothetical, but many organizations are looking into how dreams could be analyzed using more technology than just dream journals and interview analysis. We can already record brain activity and detect when people enter REM sleep. It's only a matter of time before we can capture dreams as they occur.

From there, it's just one scary step to inserting dream "programs" into people, rather like the movie "Total Recall." It would be the ultimate escapism.

I wonder if an AI of tomorrow would dream…

# # #

SUMMER READING

Artificial intelligence, blockchain, and new music technology

June 1, 2018

The summer rains arrived early this year, which means it's time to select a few good books for reading by the pool or taking to the beach. Out of the five titles I've chosen this year, three of them are about artificial intelligence and robots, one of them is about blockchain technology, and one of them is about the new world of music.

The first book is "Rise of the Robots: Technology and the Threat of a Jobless Future" by Martin Ford (Basic Books, 2016). This is a very readable summary of some of the issues facing society in the near future. I've lectured on the topic of artificial intelligence and its effects on the workforce several times, and this book provides a lot of material for discussion.

The second book is "The Fourth Age: Smart Robots, Conscious Computers, and the Future of Humanity" by Byron Reese (Atria Books, 2018). The phrase "fourth age" is related to the term "the fourth industrial revolution" put forth by the World Economic Forum in 2017, which states the "Fourth Industrial Revolution is … characterized by a range of new technologies that are fusing the physical, digital and biological worlds, impacting all disciplines, economies and industries, and even challenging ideas about what it means to be human." Heady stuff indeed.

The third book is "Robot-Proof: Higher Education in the Age of Artificial Intelligence" by Joseph Aoun (MIT Press, 2017). Education has proven to be a difficult area to disrupt, but now that many white-collar jobs can be automated, there is a renewed debate about the nature of education in a world where many of today's jobs will vanish by the time the next generation of students graduates. The book also covers the changing nature of the professor in an environment where artificially-intelligent chatbots are already replacing teaching assistants.

The fourth book is "Blockchain Revolution: How the Technology Behind Bitcoin Is Changing Money, Business, and the World" by Don Tapscott (Portfolio, 2016). Blockchain is the technology behind bitcoin and other cryptocurrencies, but it's increasingly used in other application areas. Blockchain is a revolutionary development that I believe will soon alter many areas of commerce.

The fifth book is "World In My Eyes: The Autobiography" by Richard Blade (Indigo River Publishing, 2017). The author made his name as a DJ, first in England and Europe, and later in California. He's currently one of the on-air personalities on Sirius XM, where I listen to his "Classic Alternative" show with music from the 1980s. He can do this show from his home, but still reach millions of listeners, all on a subscription basis, across multiple devices. Who needs old-school radio transmitters anymore?

# # #

iOS 12

New apps to help you stop using apps

June 8, 2018

We've all seen it: people walking around like zombies, staring intently at their smartphones, blissfully ignorant of their surroundings. It's even worse at restaurants, where people are meant to be speaking to one another, but instead, they are glued to their screens. A recent Dilbert cartoon referred to this phenomenon as "smartphone syndrome," where "One hundred percent of your mental and physical problems are caused by using your phone too much."

At this week's Worldwide Developer Conference (WWDC) in San Jose, Apple acknowledged this problem, which to many therapists verges on addiction. It seems even Tim Cook, Apple's CEO, was alarmed when he realized how much time he was spending on his iPhone every day. But in the technology world, where there's a problem, there's a solution. In this case, the solution is an app to help you stop using apps. Really.

As part of the rollout of iOS 12, the iPhone's latest operating system, Apple is providing Screen Time, which they describe as "New tools [to] empower you to understand and make choices about how much time you spend using apps and websites." One of the more interesting parts of Screen Time is "Activity Reports," which provide details on exactly how much time you've spent in each app on your phone. This includes how often you receive notifications, how many times you check your email, and warnings that you may be nearing your set limit for the day.

I can see how these features would be of great interest to parents concerned about the amount of time their teenagers spend on Instagram. But the irony here is, to provide these detailed usage reports, Apple must track and gather all your usage data. In other words, if there was any doubt that your every action was not already recorded (and possibly sold to advertisers), those doubts are gone.

I found most of the rest of the updates to iOS 12 to be relatively minor. I had the same feeling for the new Mac OS, called Mojave. There are a few enhancements, but most of the new features are incremental improvements and tweaks to the existing system. Regarding innovation, it seems the pace has slowed considerably.

Apple keeps pushing augmented reality (AR) to their developers in the hopes that more apps will be created to take advantage of this relatively new technology. But I've yet to see anyone clamoring to use AR on a daily basis. Some interesting educational offerings provide an enriched experience for youngsters and students, and of course some gamers like the immersive environment, but as of yet I don't see much mainstream demand.

# # #

WORLD CUP

Predicting the winner with data science

June 15, 2018

The World Cup starts in Russia this week. Who will win?

Several countries are perennial favorites, and the Las Vegas bookmakers show them as likely winners again this year: Brazil (4/1) and Germany (5/1). Spain was given 6/1 odds, but since they just fired their head coach a few days before the start of the tournament, I'm not sure they'll still be in the running.

Vegas is a great place to go if you're interested in gambling. If your interests run a bit more to the mystical side, you can consult the Russian cat Achilles, purported to be a psychic in predicting match winners. Achilles has already predicted host Russia will beat Saudi Arabia in the first game. Given a choice, I guess I prefer the forecasts from a cat when compared to the predictions made by Paul the octopus in 2010.

Bookies and magic animals are all well and good, but today we have data science to help us predict the results of events. It didn't work out so well for Brexit or the US election in 2016, but that hasn't stopped technology outfits from weighing in. Similar techniques are already successfully employed in other sports, in particular, baseball, to great effect. The Houston Astros are perhaps the best example of a team that is extremely data-driven – and the results have been very impressive. The players are said to have been apprehensive at first, since some of the actions dictated by algorithms seem to go against common sense, but they have proven to be effective.

Many analytics companies are using the opportunity of the World Cup to demonstrate the power of their predictive tools by crunching the numbers and predicting who will be the champions when the final match is played on July 15 in Moscow. The machine learning algorithms use input data representing artifacts such as home and away wins for each team, individual player performance, weather, travel, national league results, and even health reports, to produce an overall ranking of teams and estimated goal spreads for each game.

Goldman Sachs is reported to have used their artificial intelligence programs to run over two million possible model outcomes. They predict that the final will be between Brazil and Germany, and that Brazil will emerge victorious.

One of the fun things about sports is that they are inherently unpredictable. Predictions are interesting and can be quite accurate, but in the end, it comes down to 11 players on each team, and how they perform during the game. Everything else that happened beforehand is irrelevant.

Personally, I'm putting all my money on Iceland.

# # #

Watson Wins Bigly

Machines learn the art of persuasion

June 22, 2018

Scott Adam's most recent book, "Win Bigly: Persuasion in a World Where Facts Don't Matter," describes how some people are better able to convince the public to accept certain points of view than others. Adam's called such people "master persuaders," and he uses President Trump as an example, relying on vignettes from the 2016 presidential election to illustrate why candidate Trump's message resonated with so many people – even when it was shown that a lot of his speech was full of hyperbole and semi-truths.

Clearly, learning the art of persuasion is something of great value in today's media-saturated environment. If objective facts are reduced to subjective opinions, being able to convince others to act the way you want them to act is an incredibly valuable skill. Politicians try to do this all the time, but everyone could benefit from more insight into what it takes to win an argument when logic's role is diminished.

Can a machine learn the art of persuasion?

Yes. In fact, it's already been done. IBM's Debater project made its public debut recently. Debater is basically IBM's Watson supercomputer, trained to carry on debates in the traditional manner. The Debater project has been underway since 2012, led by IBM's Haifa team in Israel, building upon Watson's success in Jeopardy! in 2011.

In San Francisco, a professional debater argued for (and against) two topics: government support of space exploration, and increased investment in telemedicine. IBM's computer was the other debater. The format was opening statement, rebuttal, and closing remarks, all within four minutes. The Debater computer spoke in a natural voice and responded in real time to its human challenger. It was an impressive performance.

Debating is the ultimate form of persuasion, and IBM has now shown that a machine with artificial intelligence is fully capable of convincing an audience of the merits of its position. Debater will be delivered as a cloud service, so you can imagine a future version of Alexa, augmented with debater skills, carrying on ever more complex conversations with you. I could even see spouses using Watson to tune their arguments at home for maximum effect.

Indeed, imagine Debater partnering with Google Duplex (conversation) and Facebook (fake news), and soon we'll have an AI that can argue with you about anything – including why certain articles should (or should not) be believed. Politicians will spar with the computer to hone their message to be the most persuasive – facts notwithstanding.

I can't wait until the 2020 election cycle. Maybe we'll see human candidates debating computers as proxies for other candidates. We're rushing into a post-truth world and technology is leading the way.

# # #

MAKE TECHNOLOGY GREAT AGAIN

We can do it

June 29, 2018

My 21st book was just published by CTS Press. It's titled "Make Technology Great Again" and is available in print and Kindle formats from Amazon.com. It would make a perfectly patriotic gift for July 4.

President Trump's campaign motto, "Make America Great Again," inspired the book's title because technology also needs to be made great again. We're often baffled by the behavior of the electronic devices in our lives. Technology today has become too complicated, too insecure, and too unreliable.

Complexity: Do you remember when you could actually repair something that broke in your home? That time has passed, in part, because the economics of repair versus replace have swung towards replacement nearly all the time. That's because most devices are so complicated that almost no one understands how they work. Look around for anything electronic and pretend a child asked you to explain it. Everything from a malfunctioning TV remote to a blinking clock you can't reprogram on the stove is basically a black box of mystery.

Not too long ago, if your car stopped working, popping the hood and taking a look at the engine might offer a few clues to the problem because you could see mechanical parts. Today, you'd need a circuit board analyzer and an oscilloscope to figure out what went wrong. Modern automobiles are basically very complex mobile computers. They are supercharged with new chips, not new intakes.

Security: Complexity is the bane of security, and with many of our devices connected to the Internet, hacking has become widespread. Computer companies push the onus on the user to keep their machines updated with weekly patches, but few people follow the instructions correctly. How do you even know if your antivirus program is working correctly? You can't. It's even worse for the enterprise, where the number of machines makes the scale of the problem virtually unmanageable.

You've probably heard of cryptocurrencies like bitcoin. Are they a safe investment? To answer that question, you'd have to understand the security features of blockchain, a technology for a global, distributed ledger system. Good luck with that.

Reliability: Complex and insecure systems tend to be unreliable. In systems engineering, there is a philosophy of graceful degradation. This means that when a system starts to fail, it should do so in an orderly manner. But today's complex systems rarely follow this path; instead, they fail catastrophically.

Consider a major airport like Atlanta. A few thunderstorms in the area leads to a few delays on the ground, which quickly leads to total gridlock on the Eastern seaboard. The system is not as resilient as it should be. This needs to change. We can do it.

# # #

CUTTING THE CORD (PART 3)

Life with Hulu

July 6, 2018

QUÉBEC CITY – I've lived without cable television for over four years now. I've written twice about my experiences with cutting the cord in this column. In general, life is still good. However, every once in a while, something happens that makes me rethink my television options. This year, it was the World Cup.

The World Cup matches are broadcast in the US on Fox and FS1 (which is not available over the air). Moreover, several matches during the opening round take place in parallel, which means hopping between stations or recording the games for later viewing. But recording shows usually means stepping back in time, to an era of set-top boxes and DVRs. That was not something I was prepared to do.

Fortunately, technology has changed quite a bit since the 2014 World Cup. Now we have several Internet-only TV services to choose from. I examined most of the major players and decided to go with Hulu. I chose the "Hulu with Live TV" plan, which costs $39.99 per month (plus taxes).

There are many positive aspects of Hulu. This plan lets you watch most of the major networks where you live, plus many of the cable stations. It also gives you a cloud-based DVR facility, which means you can record shows for later playback, all without any cable boxes or other equipment installed in your home. Everything with Hulu is done through an app or on their website.

Unfortunately, there are several negative aspects of Hulu as well. Some of these are not strictly Hulu's fault. For example, you can only watch live TV from near where you live. Hulu uses geofencing technology to enforce this licensing limitation. It also means that when I'm traveling abroad, as I am now, I can't access most of the Hulu content, since it's restricted to US broadcasts. This means I may miss the quarterfinal matches of the World Cup on Friday and Saturday – the very reason I purchased Hulu in the first place.

Hulu's user interface is also in need of some work. Sometimes it's nearly impossible to find shows you want to record – like upcoming World Cup matches. When you do find them, it won't let you set the recording more than a day in advance. So, I could not set the system to record the matches I wanted to see before I left home.

By far, the most annoying part of Hulu is advertising. Even with the most expensive package, you are still forced to watch advertisements. The system won't let you fast forward over them, even in the recorded programs. For this reason alone, when the World Cup end, I'll probably cancel Hulu.

# # #

WORLD CUP – VIDEO ASSISTANT REFEREE

Improving sportsmanship with video technology

July 13, 2018

The 2018 World Cup ends in a few days. Were your predictions correct? My pick, Iceland, crashed out early. But so did perennial powerhouses Germany, Argentina, and Brazil. Now it's down to France versus Croatia. I doubt many analytic models had picked these two teams for the finals.

I've enjoyed watching the matches this year. One change to the game that's made an impressive difference is the introduction of video assistant referees (VAR). These are FIFA referees stationed in a broadcast center in Moscow. They are called upon to provide additional insight in near real-time to the officials on the ground when there are questions about a call on the field.

The technology used to support VAR is quite sophisticated. Fiber optic cables connect the remote officials to stadiums across Russia, and directly to the match referees – they can communicate via earpiece with the officials on the pitch at any time.

There are multiple video feeds from games all over the country that connect to the VAR control center. These feeds provide numerous camera angles for all games. If one of the on-site referees or assistants think a call was questionable, the referee makes the sign of a rectangle, which is meant to indicate a television screen, and requests a VAR review.

The VARs can use goal line cameras to carefully examine whether or not the ball crossed into the net, or if a player interfered with the goalie or a member of the opposition. The same technology is used to review offside calls. Goal line disputes and offsides are two of the most common sources of disagreement in soccer, so the VAR provides a useful and objective perspective. During play this year, several calls were reversed based on VAR analysis.

I've found that VARs have reduced one of the more annoying aspects of international soccer: faking injuries. In the past, many players would react to a slight nudge by falling spectacularly and flopping around on the turf like a fish out of water, grimacing and grasping their shins as if they were in extreme agony. Amazingly, as soon as they were awarded a free kick or a penalty shot by the referee who mistakenly bought their act, the so-called injured player would jump up and run around as if nothing had happened. There were fewer such episodes this year (Neymar's antics notwithstanding) because players know their every action is being recorded. If a VAR review is requested for a player diving, they can be penalized much more easily than in the past. Technology has helped improve the honesty and sportsmanship of the beautiful game.

#

AI AND JOBS

Automation is taking jobs from both ends of the spectrum

July 20, 2018

Artificial intelligence (AI) continues to dominate the technology news. AI startups are blossoming like desert flowers after a spring rain. Venture capital continues to flow into these startups at a fantastic rate. There's a lot of money being made, but what is the impact of AI on the economy as a whole? More importantly, what is the effect of increased automation on people's jobs?

In the past, automation tended to affect blue-collar jobs, such as factory workers. This continues to be true, but the reach of AI is growing, which means even more manual labor is destined for automation. This will have a profound impact on our society, as those struggling on the bottom rungs of the career ladder see the path upwards cut off by robots and algorithms.

Consider oil field workers. For many years, roughnecks could get a good paying job with just a high school diploma. With the current shale and gas boom in places like Oklahoma, Texas, and Wyoming, there are still opportunities for manual workers to earn $150,000. However, those opportunities are far fewer than they were just a few years ago, and they are shrinking rapidly.

A recent article in the *Wall Street Journal* chronicled how AI and automation are affecting oil and gas production. Even though we're producing about 10 million barrels of oil a day – more than ever before – employment in the industry is down 21% since 2014. The oil companies are doing more with less through the use of AI and automation. Jobs like well logger are gone, replaced by sensors and technology.

At the same time, white-collar workers felt sheltered from the growing power of AI because there was an assumption that most jobs involving human interaction and creative thinking could never be automated. That assumption is proving to be false.

In January 2014, I wrote a column here called "Predicting Literary Success." The column asked whether or not computer algorithms might replace agent expertise when it comes to picking literary winners. The answer was "yes."

I recently learned about an AI program called ScriptBook from a Belgium-based company that analyzes screenplays to determine which will be a box-office success and which will be a flop. It's impressively accurate. If you were a movie studio, you'd be very interested in using such a system.

If you were a script reader, a development executive, or a market researcher, you probably wouldn't like the system at all, because it will obviate your job.

AI and automation are taking over roles that previously were thought to be untouchable: artistic, creative, white-collar jobs that no computer could ever replace. Except now they can.

# # #

ROBOCOP RETURNS

Automated law enforcement is coming soon

July 27, 2018

The 1987 film "RoboCop" by director Paul Verhoeven and starring Peter Weller was set in a future version of Detroit so overwhelmed by violent crime that the government decided to privatize the police force. The corporation that won the contract, Omni Consumer Products, used robots to keep the peace.

However, because society would not accept fully autonomous law enforcement officers, Omni introduced cyborgs, which are a blend of a robotic body with a human brain. The storyline is similar to that of the 2017 movie "Ghost in the Shell" with Scarlett Johansson. The same problem occurred in both films: the human brain begins to remember its past life and rebels against its controllers. Things go downhill from there.

Now, RoboCop is returning. Neil Blomkamp, who directed the film "District 9" about aliens living in South African slums, is working on a sequel that ignores the 2014 RoboCop remake. Time will tell how the new movie incorporates advances in technology that have been made in the last 30 years.

Are we ready for automated law enforcement?

It's not as far-fetched as you might think.

Aren't speed cameras that photograph cars as they cross intersections a form of automated law enforcement? If they detect a car speeding, they issue a ticket. No police officer is involved. They are controversial, but their use is widespread.

Moreover, we already have drones flying around the globe, operating as part of the military and as part of homeland security. Human operators usually control the drones, but technically there's little reason to keep humans in the loop. Artificial intelligence has advanced so much that drone swarms that exhibit autonomous behavior are already possible.

When people talk about self-driving cars, we usually think of taxi service or big-rig trucks. Why not self-driving police cars for surveillance and RoboCop for enforcement? They could enter no-go zones that are challenging for current police forces. They could replace the old beat cop with a formidable presence in the most dangerous of neighborhoods. They could also make terrible mistakes if their programming was incorrect – but people are prone to errors as well.

We tend to think of RoboCop as a giant machine, but there's no reason why it could not be quite small, somewhat like Ant Man. Small robots have proven to be capable of doing everything from self-assembly to drone pollination. They can operate on land, in the sea, in the air, and soon, in space. Given the terrible toll conflicts such as those in Afghanistan and Iraq have taken on our military personnel, would we be prepared to outsource some of those duties to RoboCop? Maybe.

# # #

VIDEO KILLED THE RADIO STAR

Streaming and the Internet killed the music star

August 3, 2018

Last Friday evening I had the pleasure of attending a concert by the legendary rock band "Yes" in Orlando. They are touring to support their 50th anniversary as a group. They've gone through numerous personnel changes over the years, but their essential sound remains the same. I've seen them live numerous times and they still put on a fantastic show.

The current keyboardist for Yes is Geoff Downes. He first joined the band for their 1980 album "Drama," which featured Trevor Horn on vocals. Horn was replacing Jon Anderson, whose high-pitched voice had become a signature of Yes' progressive music. Drama was my first introduction to Yes, so the addition of Downes and Horn was not as disruptive for me as it was for other fans who had followed the band since their debut in 1968. I grew to know Yes from songs like "Machine Messiah" and "Tempus Fugit."

However, when Downes and Horn joined Yes, they were already famous for being the duo known as "The Buggles." When MTV first went live on August 1, 1981, the very first video ever played was "Video Killed the Radio Star" by The Buggles. It was a prophetic video, because for most of the 1980s videos ruled. The musical medium took second place to the visual space. Videos did have a profound effect on radio and how bands marketed themselves. How you looked was just as important as how you sounded.

Nearly 40 years later, we're going through another technological shift that is shaking the foundations of the music industry, and it's even more profound than the introduction of Napster and widespread piracy from 20 years ago: streaming.

Very few people buy actual CDs anymore – they listen to music using streaming services. I do it, too. There is a virtually infinite library of music available for us to play at any time. It's extremely convenient, and the cost is not onerous. As long as you don't mind not owning the music in physical form, streaming is great for the consumer.

However, for the musicians, streaming has severely impacted their royalties. Individual profits are way down. This is one of the reasons older bands like Yes are still touring: it's one of the few ways left for them to make money. Like popcorn at the movie theater, selling t-shirts to eager fans can be surprisingly lucrative.

It's true that video killed the radio star, but it's also true that streaming and the Internet killed the music star. Even the big recording companies are no longer in charge. Music is now in control of the tech giants. Machine Messiah indeed.

# # #

SOCIAL SILOS

Knowledgeable humans versus recommender systems

August 10, 2018

Last week I mentioned that the 1980 album "Drama" was my first introduction to the progressive rock band "Yes." Songs like "Into the Lens" really resonated with me and made me want to learn more about this band that had been putting out music since 1968. So, I did what anyone back in the day would do: I went to my favorite used record store and browsed the stacks of milk cartons full of vinyl.

The store was called "Cheap Thrills." It was located a short walk from my university, which meant I made nearly daily trips to see what new gems might have arrived overnight. I grew to know the knowledgeable salespeople, and they grew to know me and my evolving musical tastes. I spent a lot of time in the store – and a lot of money.

I was able to build up my music library of Yes records. I soon discovered that their previous albums sounded very different from Drama. However, I grew to like them as well. When Yes released their follow-up album "90125" in November of 1983, they sounded different yet again. Their original singer had returned, but their original guitarist was replaced by a newcomer. Some people were turned off by the more commercial sound, but I embraced it.

Today, record stores are a thing of the past. I sold my vinyl long ago. Most of my CDs are sitting in storage. I listen to music via streaming services. More importantly, I rely on algorithms to curate song lists for me and to recommend music that I might like. I've given up the human touch of knowledgeable salespeople for the cold logic of recommender systems.

The same phenomenon is at work on social media, where many people now consume their news (real or fake). We are increasingly allowing ourselves to be placed in silos, where everything available to you is similar to everything else. That's how recommender systems work: if you like X, then you'll probably like Y. An unfortunate result is that you'll never be able to experience Z because it's too far outside your comfort zone. I like Yes music, but I also like reggae music. How would the machine ever know this? And how would I know I liked reggae if I never had a chance to hear it?

The broader issue with this technology is that we are being cut off from one another. We are living in electronic echo chambers, where everyone says, likes, and does the same thing. If someone has a different opinion, they are derided in the most vicious manner possible. How is this progress?

###

BLOCKCHAIN

It's about a lot more than Bitcoins

August 17, 2018

Bitcoin is the digital currency that was on a roller-coast ride until the end of 2017. It's since settled down somewhat but investing in Bitcoins remains fraught with risk. This is also true for other cryptocurrencies such as Ethereum and Litecoin.

What makes Bitcoin possible is an underlying technology known as blockchain. A blockchain is made up of a series of data blocks, which are recordings of transactions like an old-fashioned ledger. These blocks are connected forming a chain. Unlike a single ledger book, a blockchain is distributed across a network, such as the Internet. Computers on the network that are part of the blockchain are called nodes. They act as both consumers and providers of data to their peers, like a BitTorrent network.

The attraction of blockchain is that it facilitates the transfer of assets, such as digital money, between two parties that don't have to trust one another. Traditionally, if you want to send money to someone, you pay by credit card or use a wire transfer. Both mechanisms require a trusted third party, such as a bank or a securities clearing house, to act as an intermediary. This adds cost to the transaction, and it takes time, which is one of the reasons it takes 3-4 days for stock market transactions to settle. With blockchain, the transaction can be completed securely and verifiably in a matter of minutes.

The security in a blockchain is provided by special nodes on the network called "miners." These nodes compete with one another for financial rewards (e.g., new Bitcoins) and to be the node that is allowed to update the blockchain with the record of a transaction. They compete by solving increasingly difficult mathematical problems. There are now thousands of computing nodes in the Bitcoin blockchain that burn through enormous amounts of electricity to run special-purpose hardware that is optimized to solve cryptographic puzzles.

A blockchain is also transparent, in that all nodes can see every transaction. A copy of the entire blockchain is held at miner nodes. This chain can become quite large. According to Statista.com, the size of the Bitcoin blockchain has been growing since the virtual currency was created in 2009, reaching approximately 173 gigabytes by the end of June 2018.

Some of the most interesting uses of the blockchain are in areas other than Bitcoin. For example, in government, voting records can be kept in a blockchain that everyone can see, but no one can hack. Or in real estate, where land transfers can be recorded in a blockchain to speed up the closing process. Blockchain has the potential to be a truly transformative technology.

# # #

BACK TO SCHOOL

But not for me — I've retired

August 24, 2018

Students all over the country are in denial that summer is over. Their parents, however, are probably delighted. That's because it's back to school time. Classes have already begun for most districts and universities. It's time to leave the beach and hit the books.

But not for me. For the first time in 20 years, I'm not heading back to school. I've retired as a professor emeritus from the Florida Institute of Technology. I've enjoyed my time at FIT since I joined in January 2003. I've had the opportunity to work with many exceptional individuals. It's been a great ride.

I'm not retiring from professional life. I'll remain active in groups such as Big Data Florida, the Center for Technology & Society, and INCOSE. I have several book projects underway. And of course, I'll still be here, writing this column.

While packing my office, I felt as if I was taking a walk down technology memory lane. Some of my books and journals go back to the time I worked at IBM in the late 1980s. These are mostly books on programming languages and compilers, but there were also a few on technical communication.

There were several shelves of books from my doctoral studies. These are more about theoretical computer science, topics I haven't really studied or taught since I graduated in 1995. But reading all the notes I made while studying these subjects in preparation for the grueling comprehensive exams brought back many details in a great rush of wonder and dread. Wonder, that there was so much interesting material to read. Dread, because I didn't know how I'd ever stuff all that information into my head in time to pass the exams. Thankfully, things worked out.

The vast majority of books were purchased during my time as a research scientist at Carnegie Mellon University's Software Engineering Institute, and as a faculty member at the University of California, Riverside, and FIT. The books are topically clustered in time. For example, there was a shelf full of books related to SQL and Perl programming for funded research projects with BMW and other companies.

Most of my books are about software engineering, reflecting the majority of my time at FIT. More recent titles concerned educational technology, systems engineering, and data science. It's incredible how each book represents a point in time during a long career.

I'm confident that Florida Tech has only bigger and better days ahead of it. The phrase "high tech with a human touch" has never had more importance than now. But for me, it's time to put on the white belt, black socks, and beach sandals.

# # #

A Medical Comedy of Errors

Complex systems fail when you need them the most

August 31, 2018

Stop me if you've heard this one before. You're in a rush to get something done, and every bit of technology you touch fails you. What should have been a relatively simple task turns into a comedy of errors – except it's not very funny. Sound familiar?

I've recently been "enjoying" the administrative experience of changing benefit plans. Considering we live in a digital world, there's an impressive amount of actual paperwork involved. I was already frustrated because I have a looming deadline to complete the change process, and technology was getting in the way.

When I tried to login to the main website of my benefits coordinator, the system locked me out. It said I had tried too many times to login unsuccessfully. I had not, but I was forced to call them to reset my password. And by "them" I mean an automated system, not a person.

The last message I heard was that my password had been reset, which was good to know, except the system didn't tell me what the new password was, so I still couldn't login. I called again, was switched to someone who promptly hung up on me during a transfer. This was after entering my phone number when I called and then being asked for the exact same information by the next person on the phone. Have we not mastered the technology of caller ID in these systems yet?

I decided to postpone dealing with this organization and try something simpler. I needed to obtain proof of current dental coverage. So, I tried to login to my dental insurance carrier. You probably know what happened next: my password had expired. My new password link arrived by email – several hours later.

When I did login to the dental system's website, I tried to print my ID card. The first time I tried, a new tab opened in my browser with a blank screen. In the middle was a tiny message: "Plug-in blocked." It didn't say which plug-in, so I assumed it was for Flash. I enabled it, reloaded the page, and saw the same error message. Why are these systems so terrible at informing users of the underlying problem?

I opened a different browser and repeated the login sequence. This time, there was no plug-in error. Instead, the dental system's website kindly reported a "500 server error," and told me, "Relax. You didn't do anything wrong. The page you requested is temporarily unavailable." This is not rocket science; this is just displaying an ID card.

I wonder if the CEO of these companies ever try to use their own systems?

# # #

FLORIDA TECHNOLOGY

Living the good life in the sunshine state

September 7, 2018

I just published my 22nd book, called "FLORIDA." It's available now on Amazon.com in both print and Kindle formats. It's a collection of stories about living the good life in the Sunshine State.

If you ask people who live elsewhere to describe Florida, they tend to use words like sunshine, hurricanes, Disney, alligators, space, pythons, beaches, and retirement. These words are evocative of modern Florida: endless summer and destructive storms, amusement parks and deadly predators, invasive species and gateways to the final frontier. But there's much more to Florida today – there's technology. Lots of it.

We are fortunate to live on the Space Coast, a unique area of east-central Florida that is a narrow strip of Atlantic shoreline. This is where the nation's space program began – and continues to thrive. I've seen Space Shuttle launches and landings, huge rockets taking off, and experienced the remarkable scene of SpaceX's Falcon 9 boosters returning to earth. It's the only place I know where local hotels put three items on their beachside information boards: the air temperature, the ocean temperature, and the next launch time.

Several of the stories in "FLORIDA" focus on the space program, but there are many other types of technology that drive our state's economy. Consider the numerous theme parks in Orlando. They provide unparalleled entertainment for millions of visitors each year. But the engine making all those rides and exhibits possible is technology. Many of my past students are employed as software engineers at places such as Disney and Universal, working behind the scenes to make the magic possible.

Brevard County is also home to a large number of nimble manufacturing facilities. They can switch production lines to accommodate different orders from multiple customers due to the flexibility of their control systems. Many hardware and mechanical engineers provide the innovation to make local manufacturing a thriving enterprise.

Communications systems are the bread-and-butter of several local companies. For example, they provide radio systems for law enforcement and the military to operate in the hostile environments around the globe. They also offer sophisticated systems to work in space, which powers both our aerospace and defense industries.

There's been a surge in companies across the state that have developed a reputation for excellence in cybersecurity. Many of these companies are the result of fundamental research at our leading universities that spin-off the results to commercialize their ideas. The nature of cybersecurity means that you don't often hear much about these companies, but they are there, working in the background for our benefit.

Modern Florida is more than citrus, swamps, and mosquitos. It's about living the good life – powered by technology.

# # #

COMPUTERS AND BUGS

Which came first?

September 14, 2018

Which came first, the chicken or the egg?

This philosophical dilemma has vexed people for years. A chicken is hatched from an egg. But the egg is produced by a chicken. It's this sort of circular logic that gives graduate students nightmares.

There's a similar problem related to software. Which came first, the computer or the bug? Given the sorry state of many applications, the answer should not be surprising: the bug. Shortly after the first bug was discovered, scientists formed the first computing association. Rather like closing the barn door after the horse has escaped.

The word "bug" in a computer program means there's something wrong. The origins of the term come from the first recorded instance of a literal bug in a program. On September 9, 1945, Grace Hopper made an entry in her notebook that the source of the error in the Mark II computer at Harvard had been found: it was a moth that was stuck between two relays. The insect's body was causing a break in the circuit. This was well before the days of vacuum tubes, never mind transistors and integrated circuits. The computer used physical relays, rather like light switches, to perform calculations. When the moth was removed, the computer functioned properly again. In other words, the program had been "debugged."

About one week later, on September 15, 1947, the world's first (and now oldest) computing organization was formed: The

Association for Computing Machinery (ACM). The ACM, along with the IEEE Computer Society, is a leader in the computing field. As stated on their website, "With nearly 100,000 members from more than 190 countries, ACM works to advance computing as a science and a profession."

I've been a member of the ACM for decades. In fact, I was the chair of their special interest group on the design of communication (SIGDOC) for several years. I've also run numerous conferences and workshops that were sponsored by the ACM. I would encourage all computing professionals to consider membership in organizations like the ACM. They are a great way to continue your education, to network with fellow academics, enthusiasts, and practitioners, and they provide a way for you to give back to the community through various volunteer opportunities.

It's ironic that the ACM was formed shortly after the first bug was found. The two events were probably unrelated, but a historical perspective of the timing is interesting. Perhaps if the ACM had already existed, some of their best practices in software engineering could have changed the computing culture from one focused on debugging to one focused on avoiding bugs in the first place.

# # #

The End of Home Cooking

When fast food is not fast enough

September 21, 2018

Just how lazy are we these days?

Until quite recently, if you were too busy to prepare a meal at home, you went to a nice restaurant to enjoy a nice break from the kitchen. If you were in a hurry, you'd go to a fast food joint for a quick burger or pizza slice. If you were really in a hurry, you brought the food home to eat. If you were in a crazy hurry, you used the drive-through and ate the food on the way home.

Now, it seems we can't even be bothered to go to the store in the first place.

Delivery companies like Uber Eats, DoorDash, and Grubhub are all the rage. They let you order food from your favorite restaurant right from your smartphone and have the tasty dishes delivered to your home. Investors love them, and so do consumers. They are also providing traditional eat-in restaurants with a new source of income. Have you noticed all the delivery people waiting to load their carts with food for other people at the cash register?

The business model of home delivery for restaurant food is certainly not new. Pizza parlors and Chinese restaurants have been doing it forever. However, new technology has made these delivery services economically viable. Not only do they rely on Uber-like location services, they also use special food re-heating systems and advanced packaging. And they contribute to the "gig economy" by giving drivers the occasional work. But mostly they feed our need for fast food with minimum fuss.

Social scientists used to bemoan the loss of the family mealtime. It was an opportunity for everyone to sit together, enjoy some home-cooked food, and talk about the events of the day. For most modern families, such gatherings are increasingly rare. Our daily lives have now gotten so busy that we don't even have time to skip the cooking and go out to eat together. Maybe these delivery services will reverse the trend by bringing restaurant food home. Or perhaps they'll turn us all into couch potatoes, watching TV while staring at our phones, waiting for the next meal to arrive on the doorstep.

Come to think of it, maybe there's a niche market for the "last mile" of food service. Who wants to get up from the comfy chair, open the front door, lug the food inside, and laboriously unpack it? Surely there's someone who could do all that for me, and toss the crumbs into my mouth? Maybe I can hire a robot for that.

Oh, wait. I already have one. "Alexa, open an app that can order me some food."

# # #

DIGITAL BROTHELS

Human-form robots as sex workers are already here

September 28, 2018

You might want to send the kids out of the room while reading this column.

The adult entertainment industry has always been an early adopter of advanced technology. This goes back to the 1980s when VCRs first became popular. The purveyors of porn quickly realized they could close their expensive (and sleazy) X-rated movie theaters and put their product on VHS tapes instead. Sales soared.

When DVDs appeared in the 1990s, the porn industry quickly moved their product to this new format. It's not an exaggeration to say that the rapid adoption of DVD players was driven in large part due to people's desire to watch adult films at home. The same thing happened a decade later when high-definition Blu-ray discs first appeared.

However, the Blu-ray market never took off quite as planned because the porn industry moved online. Today, the proliferation of adult content on the web is every parent's nightmare. There are images, videos, and even live, interactive shows that mimic the old peep shows. And this content is not your father's dusty Playboys; it's extremely graphic, often violent, and caters to just about every fetish you can imagine – and then some.

Two recent technologies being used to make online porn more convincing are augmented and virtual reality. The latter allows the users to have simulated experiences in 3D. It feels more like the real thing, but it still only stimulates one sense: vision. The endgame for the porn industry is to make their product indistinguishable from actual sex, but to do that, you need an actual person. This is part of the reason prostitution is still flourishing. There's no other way to stimulate the rest of our senses.

Or is there?

The latest technological development where the porn industry is again taking the lead is human-form robots. These is a new take on blow-up dolls, except these robots look nearly indistinguishable from real people. The robots have AI software, so they learn to react to stimuli, such as answering questions in a natural voice. How long before society has to consider the implications of someone wanting to marry their robotic partner?

If this sounds far-fetched, it's not. Human-form robots as sex workers are already here. A company called KinkySdollS opened their first store in Toronto last year and are planning on opening their second store in Houston very soon. The company's founder said he plans to have 10 locations throughout the country by 2020. This expansion has caused a backlash with residents against "robot sex brothels." Customers pay $60 for a half-hour alone with the robot, which they can then buy at prices starting at $2,500.

Are we ready for this?

# # #

Sputnik

The space race began on October 4, 1957

October 5, 2018

I've just started a new book project called "SPACE." It will be a collection of stories about our reach for the stars. In particular, the anthology will celebrate 50 years since we landed on the moon on July 20, 1969. If you're interested in contributing to the project, please send me a note or visit http://bit.ly/2y0zRsB for full details.

We've actually had an active space program for more than 50 years. NASA was founded 60 years ago, on July 29, 1958. Before the Apollo program, we had the Mercury project. And leading up to Mercury was a series of rocket experiments. In 1926, Robert Goddard launched the first liquid-fueled rocket. During World War II, Wernher von Braun led the development of the V-2 rocket, which in 1944 reached an altitude of 109 miles, making it the first man-made object in space. The V-2 was a genuine threat to London and other allied locations near the end of the war.

But what really kicked off the start of the space race with the Soviets was the launch of the Sputnik satellite by the USSR on October 4, 1957.

Sputnik was a tiny satellite. It was shaped like a small ball, just over 22 inches in diameter. It weighed just over 180 pounds. In comparison, the Delta IV Heavy rocket can lift a payload of 63,470 pounds to low earth orbit.

Sputnik trailed four external radio antennas behind it like

tentacles on an octopus. The Soviets did not attempt to hide the satellite's journey through space. In fact, it seems clear they wanted the world to know of their achievement. Sputnik's radio signal was monitored by government agencies and amateur astronomers around the world.

Sputnik's journey only lasted for three months. After three weeks, its batteries died. It burned up on reentry on January 4, 1958. It had completed 1,440 orbits around the Earth. Its mission was an unqualified success: it announced to the world that the final frontier was now a strategic imperative for the global superpowers. In particular, it heralded the beginning of an unprecedented era of technological innovation, driven by the scientists and engineers at NASA and their contractors.

A few years after Sputnik, the USSR launched the first cosmonaut into space. Yuri Gagarin completed one orbit in the Vostok 1 spacecraft on April 12, 1961. The US followed closely behind with Alan Shepard's suborbital flight on May 5, 1961. The space race was well and truly underway.

One major impact of the USSR's early success in space was a renewed emphasis on STEM education for American children. There was a definite feeling that we were falling behind. How are we doing today?

# # #

FRANKENFOOD

Would you eat a burger grown in the lab?

October 12, 2018

As of September 2018, there are approximately 7.6 billion people on Earth. This number is expected to rise to 9.8 billion by 2050, and to a staggering 11.2 billion by 2100. These figures assume society remains much as it is today, but with a steadily decreasing fertility rate – particularly in the developed world. Most of the population growth will come from the developing world, with Africa alone contributing an additional 1.3 billion people by 2050.

How will we feed this many people?

We've made tremendous strides over the last century in increasing crop yields and optimizing land use for growing the food needed to support our current population through the use of genetic engineering. One area that has made a measurable difference is the use of advanced pesticides and genetically modified organisms (GMO). GMO food results from the introduction of DNA from another species into the original organism. This can produce crops that are resistant to certain pests, or that can grow in more arid or salty conditions. There is a lot of debate about the long-term safety issues related to GMO food, particularly in Europe. But much of the foods we currently consume are in fact GMO foods.

There is a related development in food science that is causing much discussion. This is our nascent ability to grow meat in the lab. Technically, this is not GMO, but it's not quite natural either. Synthetic meat, also known as clean or cultured meat, is made by growing muscle cells in a petri dish, where they are fed a nutrient serum, which causes the cells to develop into muscle-like tissue. If you think this sounds a lot like growing humans in a pod, as in "The Matrix," you're not alone.

Several startups are focused on scaling the process so that they can economically produce synthetic meat (and fish) products for the consumer. Very soon, you might be able to buy a synthetic burger at the local supermarket. But would you want to eat it?

Some people fear such "Frankenfoods" may have unintended side effects if consumed over the long term. These fears are shared by people who are opposed to GMO food products. Even though the process is different, the concerns are similar.

If you were a vegetarian or a vegan because you had moral objections to eating the flesh of animals, would you eat a synthetic burger? As they say, "No animal was harmed in the process." Other than stem-like cells.

#

ACM CODE OF ETHICS

The first update to guide computing professionals since 1992

October 19, 2018

The Association for Computing Machinery (ACM) is the oldest computing organization in the world. It has a tremendous influence on many aspects of computing, particularly for academics and researchers. For example, the ACM publishes numerous journals and sponsors many conferences geared towards advancing the computing field.

Computer science, and software engineering (in most places), is not a regulated profession in the same way law and medicine are governed. There are no state or federal agencies that perform licensing or provide formal certification for practitioners, which means anyone can call themselves a computer scientist, whether or not they have the necessary credentials. This may change in the future, but for now, the implication is that the discipline must provide its own set of guidelines for proper professional behavior.

The ACM just released an updated "Code of Ethics and Professional Conduct." This code is meant to be followed by all members of the ACM. It's the first time the code has been updated since 1992, when the computing landscape was dramatically different. The subtext for the 2018 version of the code is "Affirming our obligation to use our skills to benefit society," which reflects the increased influence that computing has on our lives.

The code is quite lengthy, so I selected three principles to illustrate the nature of the revised guidelines. General Ethical

Principle 1.6 says, "A computing professional should respect privacy." This is incredibly important in the age of big data and weekly data breaches. In my experience, learning how to manage user privacy properly is not something most computer scientists are taught. This has to change.

The code's second section is on professional responsibilities. Principle 2.9 says, "A computing professional should design and implement systems that are robustly and usably secure." The terms "robustly" and "usably" are important adverbs in this sentence. Many of our systems today are not robust or resilient; they are prone to failure in unexpected ways. Security is a desirable characteristic of all systems, but the way security is implemented is often at odds with user interaction. Security and usability are classic tradeoffs that engineering continue to struggle with.

The third part of the code concerns professional leadership. Principle 3.6 says, "A computing professional, especially one acting as a leader, should use care when modifying or retiring systems." This particular principle resonated with me, because I spent much of my career working with legacy systems. I'm acutely aware that there are often a multitude of hidden dependencies in these applications, and just swapping out the old for the new without fully understanding how users interact with the system on a daily basis is a recipe for disaster.

# # #

SCARY TECH STORIES

Be afraid, very afraid

October 26, 2018

Halloween is just around the corner, so it's time to share a few scary tech stories. These everyday events are so horrific, so unnatural, so alarming, that I suggest you read the rest of this column with all the lights on. You were warned.

Have you ever lost the tiny remote that comes with your Apple TV? I've done it numerous times, and whenever it happens, I realize that there's no way of turning off the device, short of unplugging it, without the remote. There are no controls on the device itself; it's just an obsidian beast that only responds to Apple magic. That's a scary form of user interface design. Time to start digging through the couch cushions.

Have you received one of those blackmail emails? They say hackers installed a virus on your computer several months ago, and they've been tracking everything you've been doing online. They also say they've been watching you through your web camera on the computer, without you ever knowing. The message includes snide comments about how you act in private and the naughty websites you visit. If you don't pay a ransom in bitcoin to someone in a distant land, they'll release photos and videos of your online activity to your friends and family. That's chilling because there's a small part of you that believes them. Everyone has secrets they don't want to share.

Have you ever left your key fob in the car? The system may warn you that the engine was switched off and the fob was left behind, but if you're in a hurry, you probably ignored the warning. The door locks automatically after a few minutes. How do you get back into the car? There is no physical key. You can see the fob lying on the car seat, but you can't reach it. If this happens at night, in a bad neighborhood, things could go wrong very quickly. Even worse: you see your phone lying beside the fob in the car. You hear footsteps coming up behind you. There's nowhere to run. You start shaking.

Have you ever forgotten your password to a website? To restore your password, the system will email you a link that you click on. But you use the same password for multiple accounts, including your email, so you're locked out of there too. You can't access the reset instructions, and you can't send an email to the help department. If you send an email from a friend's account, the help department doesn't believe you're you. You didn't setup two-phase authentication. You quickly descend into a call tree abyss, and you'll never get out. You're doomed.

# # #

Augmented Reality Dining

Le Petit Chef makes a virtual visit to your dinner plate

November 2, 2018

During a recent cruise on the Celebrity Infinity, I had the opportunity to dine at one of the ship's specialty restaurants, Qsine. This restaurant is not unique due to its food (although the bouillabaisse, lobster tail, and filet mignon were excellent). It's unique due to how the food is presented: with augmented reality (AR) in the form of Le Petit Chef.

Before each course, an animated figure appears on your tabletop. He spends a few minutes preparing the main ingredients for your next delicacy. For example, in the case of the bouillabaisse, he hauls mussels and clams from the ocean and tosses them into your bowl. He also wrestles with an octopus before slicing off a few tentacles for your soup. He disappears into the ocean on a jet ski, complete with realistic sound effects. As soon as he leaves, your waiter appears and places the real bowl of bouillabaisse in front of you. This performance is repeated for the next three courses — animation working in concert with servers in a veritable symphony of food and technology.

What's so amazing is that this 3-D effect is accomplished without the diners wearing any form of dorky eyewear. No goggles or headsets are needed. The animation appears very real in front of your eyes, overlapping with the plates and cutlery on the table. I was so impressed with the whole show that I went to the restaurant twice, to confirm that the process was repeated. It was.

The company behind the 3-D animation is a Belgium startup called Skull Mapping. They have used their technology to include adventures of Le Petit Chef in high-end restaurants around the world, in places like London and Dubai. The hologram effect is an optical illusion, created by software and special-purpose Panasonic cameras. The result is extremely effective.

Augmented reality is the focus of a lot of venture capital. Apple is all-in for AR with their iPhones and the developer's ARKit. Florida-based Magic Leap is a big player in this emerging field. But all these systems require the user to look through a screen or wear some sort of head-mounted display. Skull Mapping's product avoids all this, and in this sense, I think they are pointing the way forward to the rapid integration of AR into other parts of our lives.

During your next trip to Publix to buy some fish, instead of seeing a row of cold filets just lying on ice, imagine you could see the fish swimming inside virtual aquariums in the display case. You'd make trips to the store just to see the show – and then you'd buy even more stuff you had not planned on.

# # #

SILENT DISCO

Dancing alone together

November 9, 2018

My recent cruise on the Celebrity Infinity was full of firsts. I wrote last week about my first experience with augmented reality dining. I find myself looking for Le Petit Chef at all of my meals now.

After dinner, I experienced another first: silent disco.

Most cruise ships have a bar with a large dance floor somewhere on the upper decks so that passengers can enjoy the view while they make fools of themselves. This cruise was no different in that respect, except that the level of foolishness was taken to a whole new height.

A typical disco blares music so loudly that you can't speak to anyone in the room; you need to yell all the time. The disco on Infinity was the total opposite: silent. There was no music playing that you could hear, but people were still dancing. That's because everyone was wearing headphones, and they could hear the music perfectly well. But if you just looked across the dance floor, the only sound you could hear was the low-volume mumbling of people trying to recite the lyrics to the songs they were listening to, with the occasional yell of triumph for busting a particular dance move. Think "Saturday Night Fever."

As odd as this silent disco felt, it wasn't the oddest part of the night.

Everyone was wearing headphones, but they weren't listening to the same music. Each headphone had a switch that let the dancer select one of three different channels: 70s, 80s, and everything else. Depending on which channel was selected, a different color light was glowing on the headphones: red, blue, and green. This means you are trying to dance with your partner, which for some of us is already a significant challenge, but they are dancing to a different tune. Literally. It was actually great fun, and incredibly amusing to watch people doing their own thing, limbs flailing to whatever song they were dancing to, oblivious to anything else going on.

In many ways, silent disco is a perfect metaphor for modern society. Technology has made it possible for us to experience highly-personalized media, blissfully unaware of what other people are doing. It makes social separation more possible than ever. Think of a bunch of people sitting in a restaurant. They are all staring at their phones, each doing something different, but they're doing it together.

Although I was skeptical of the silent disco premise, I can't argue with the results. Everyone on the dance floor was grinning like an idiot, myself included, relaxed in the certainty that they'll never see these people again. Why not enjoy your connected solitude and dance the night away?

# # #

World Diabetes Day

A global epidemic with no cure in sight

November 16, 2018

November 14 is World Diabetes Day. It's sponsored by the International Diabetes Federation to raise awareness of diabetes worldwide. It's not hyperbole to say that diabetes has become an epidemic in some communities, and the complications that result from the disease can be deadly.

I was diagnosed as a Type I diabetic in October 1993. I was traveling at the time, and I felt lethargic, I was always thirsty, and I was having trouble concentrating. When I returned home, I went for a checkup, and the doctor confirmed my condition. From that day forward, I've been injecting myself with insulin every day. Before I switched to an insulin pump in 2011, I calculated that I had given myself about 30,000 needles. That's a lot of jabbing.

In Type I diabetes, your pancreas no longer produces insulin. Without it, the sugar levels in your bloodstream start to climb dangerously high. Until Banting and Best uncovered the role of insulin in the body in 1922, Type I diabetes was literally a death sentence. Now I'm alive due to synthetic insulin produced through a recombinant DNA process.

Type II diabetes is by far the more common form: about 95% of the population with diabetes have Type II. With Type II, your body still produces insulin, but your cells are unable to process it correctly. Sometimes a change in diet and exercise regimen can solve the problem. Other times, medication is needed to help your cells process insulin. A few years ago, I was told I also had Type II diabetes, which makes my condition quite rare.

There is no cure for diabetes. It's a chronic condition that must be managed for life. This seems odd, given all the modern medical technology we have at our disposal. We can insert heart pacemakers, perform liver transplants, even adapt to bionic limbs, but coming up with a replacement for the islets that produce insulin in the pancreas appears to be out of reach for now. There is something about the pancreas that makes it difficult to fix, which is part of the reason pancreatic cancer remains so deadly.

For now, I manage my diabetes through a combination of technologies. I use a blood test meter to measure my glucose levels at least five times a day. I use a pump that is attached to my abdomen to deliver frequent doses of insulin throughout the day. And now I also use a continuous monitoring device that measures my subcutaneous blood sugar levels to help the insulin pump work more effectively.

I'm basically a man with cyborg-like implants. Like millions of other diabetics, I'd much prefer a new pancreas – or the robotic equivalent.

# # #

HOW TO BECOME A DATA SCIENTIST

Prepare for a 21ˢᵗ century career

November 23, 2018

The dictionary definition of data scientist is "a person employed to analyze and interpret complex digital data, such as the usage statistics of a website, especially in order to assist a business in its decision-making." Data scientist is one of the hottest professions in the world. As a data scientist, you'll be in high demand, earn an excellent salary, and work on some of the most interesting problems facing society today. Sounds great! So, what's the catch?

It's not easy. But nothing worthwhile ever is.

Do you need to be good at math? (Some of it, yes.) Do you need to be a good programmer? (It helps.) Do you need to be a good communicator? (Absolutely.) Fortunately, these are all skills you can learn.

At the Big Data Florida meeting this week, Tauhida Parveen spoke about what it takes to become a data scientist. Dr. Parveen is lead instructor at Thinkful, a NYC-based startup focused on online education for tomorrow's developers. She leads the development of curriculum in two areas: algorithms & data structures, and data science.

In her view, a data scientist is someone who straddles three areas: mathematics and statistics, software engineering, and communication. There are related titles, such as data analyst and data engineer, but the data scientist is at the core. Their role is to convert raw data into actionable intelligence.

Tauhida described four kinds of data scientist: (1) the super analyst; (2) the statistician; (3) the AI guru; and (4) the researcher. The super analyst is the foot soldier of the data science revolution. They are expert problem solvers, great communicators, and love to build products. The statistician understands probabilities and statistical models, are math whizzes, and know how to design experiments. The AI guru lives and breathes automation, is proficient in programming, and has mastered machine learning algorithms. The researcher is searching for the next big thing, typically has a doctoral degree, and works in advanced labs in academia, government, and industry.

If you want to become a data scientist, you need to think about what type of data scientist you'd like to be. You need to be comfortable with constant and rapid change. You need to accept life-long learning as part of your career advancement. And you need to become proficient in mathematics, programming, and a variety of databases, visualization tools, and application frameworks. Above all, you need to become knowledgeable about various types of machine learning: supervised, unsupervised, and reinforcement. It is through machine learning that most advancements in data science are taking place.

Happily, there are many resources to help you become a data scientist. Immerse yourself in the field, try your hand at an open-source project, and get started!

# # #

CRISPR Babies

Eugenics has arrived

November 30, 2018

I recently delivered a lecture on the subject of ethics and genetic engineering. I spoke about clones, DNA, and the CRISPR tool used to perform genome editing. Genetic engineering holds the promise of incredible medical advancements, such as personalized gene therapy. But genetic engineering also gives us the means to perform questionable medical procedures, such as creating so-called "designer babies," whereby parents someday would be able to selectively alter their unborn child's physical traits.

On Monday, "someday" became "today."

News out of China went viral with the announcement that He Jiankui, a US-educated scientist working at the Southern University of Science and Technology in Shenzhen, had performed the first actual instance of genetic engineering on twins. The baby girls, named Lulu and Nana, were born after the procedure. The genetic modification that was performed was to edit the CCR5 gene, which is a receptor that HIV uses to enter white blood cells. The editing essentially disabled the receptor, making the babies resistant to HIV infection.

That sounds wonderful in theory, but in practice, it's unclear what possible side effects may arise from the procedure. Researchers around the world were skeptical that He actually performed the editing, but He spoke at a genetic engineering conference in Hong Kong on Tuesday and provided more details of his experiments. His presentation suggests that we have indeed crossed a threshold by creating the first genetically engineered humans.

The ethical implications of performing gene editing on living beings are profound. For example, its unclear what effect modifying the CCR5 gene will have on the children – and on their children. Unlike traditional medicine, DNA modifications are passed on to future generations. It's also unclear what affects minor errors in the gene editing process may have on the girls. It could be harmless, or it could be serious; no one really knows.

More importantly, if He's work proves to be true, it opens the door to other kinds of genetic engineering using CRISPR. What is to stop prospective parents from enhancing a fetus with "desirable" traits, such as eye color, skin pigment, or physical prowess? What about enhanced mental ability?

Taken to the extreme, CRISPR babies could represent a new schism in society: the enhanced and the regular. If genetic enhancements are available for the affluent, will that leave parents with more modest income worrying that their children will always be a few steps behind? Or will genetic engineering become so commonplace that the procedure will be covered by insurance companies?

I always assumed that super AI would one day surpass us. Perhaps eugenics will do it faster.

# # #

CUTTING THE CORD (PART 4)

Acorn TV, BritBox, and CuriosityStream

December 7, 2018

As dedicated readers know, I've enjoyed life without cable television for nearly five years. I've written here about my experiences several times, most recently over the summer while trying Hulu to watch the 2018 FIFA World Cup. It seems "cutting the cord" is a perennially popular topic, given the number of queries I receive, so I decided to share more about my online viewing habits.

Way back in 2001, I became hooked on a British soap opera called EastEnders. It's been running on BBC One since 1985. I had to subscribe to BBC America to watch EastEnders, but they stopped carrying the show a few years after I started watching it. That led me (and many others) to find alternate ways of getting my daily fix of life in London's raucous East End, which usually involved dodgy proxy servers.

Last year, the BBC finally relented and made EastEnders (and much of their other content) available to third parties for licensing and rebroadcast around the world. Fortunately for me, a new online TV service had just launched: BritBox. I watch BritBox using an app I downloaded to my Apple TV. I pay a $6.99/month subscription fee, which is added to my Amazon.com Prime account. BritBox does not provide a live feed (I use the SkyNews app for that), but BritBox's library of content is quite impressive.

Through BritBox, I can finally watch all the recent episodes of EastEnders, along with other BBC shows like Question Time and

Prime Minister's Questions (to keep up with all things Brexit). I can also watch many of the excellent mystery series that the UK is so famous for, such as Midsomer Murders and Vera. And for old-school soap fans, BritBox also carries Coronation Street.

If you'd like to expand your viewing options beyond the UK, consider Acorn TV. It works much like BritBox, installed as an app, and billed through my Amazon.com account for $4.99/month. Acorn TV does carry a lot of UK content, such as the wonderful series Doc Martin. But Acorn TV offers a lot of unique content that I really enjoy. For example, you can see comedies and dramas from Australia and New Zealand – something that was nearly impossible just a short time ago. I genuinely like the show 800 Words. It provides reasonably accurate insight into the life of a newspaper columnist. (Ahem.)

If you want science and education, you can complete your add-on package with CuriosityStream. It costs $2.99/month and works just like BritBox and Acorn TV. CuriosityStream's niche is documentaries. They carry shows covering everything from medieval history to quantum computing.

Don't be afraid to cut the cord; TV can be great again.

# # #

GOOGLE'S 2018 YEAR IN SEARCH

Find out what people were looking for around the world

December 14, 2018

Google recently posted a summary of the "Year in Search" for 2018. It's an intriguing glance at search trends around the world during the past 12 months. You can see the results yourself at www.google.com/2018.

By default, the Google website displays search results for the entire world, but you can narrow the report to specific countries. It's a fascinating window into how different geographical areas were focused on different events. For example, the top search phrase for the world in 2018 was "World Cup," reflecting the global interest in FIFA's soccer extravaganza. By comparison, in Canada, the top national news search phrase was "Humboldt Broncos," a tragic event last April where sixteen people were killed and thirteen injured, most of them teenage hockey players, when their tour bus collided with a semi-trailer in Saskatchewan.

In the United States, the person most searched for was Demi Lovato. She's a singer, songwriter, and former Disney Channel actress, and I have to admit that I did not recognize her name. I'm not sure what that says about how detached I am from popular culture. Rounding out the top five people were Meghan Markle (a new member of the British royalty), Brett Kavanaugh (a new member of the Supreme Court), Logan Paul (an online video celebrity), and Khloé Kardashian (someone I can't keep up with).

There is also a search category called "Beauty Questions."

Apparently, a lot of American teenage girls are very worried about their eyelashes. The top five search phrases were:

1. How to apply magnetic lashes

2. What is a lash lift

3. How to remove individual eyelashes (ouch!)

4. What hair color looks best on me

5. How to do cat eye (I don't even want to know what that's all about)

There's an interesting search category called "What is...?" The top search for 2018 was "What is Bitcoin?" That makes sense, given the rollercoaster ride the cryptocurrency had in the last year. The second-most popular search phrase is even more intriguing: "What is racketeering?" Were that many people really following events involving international criminals and the RICO act? Rounding out the list were "What is DACA," "What is a government shutdown," and "What is Good Friday?" Given the state of the federal government these days, the answer to the shutdown question may become apparent quite soon.

In the United Kingdom, the results for the same "What is...?" category were quite different. People in the UK were likewise interested in Bitcoin, but they were also interested in GDPR, ibex, nerve agents, and the commonwealth. For the most part, these four terms have little relevance to us. Why are the British so interested in a wild goat?

# # #

LOOKING BACK AT 2018

Big data, privacy, and security; genetic engineering; cutting the cord

December 21, 2018

The winter solstice is almost upon us, which means the end of the year is fast approaching. It's time to take a look back at some of the interesting technology developments from 2018. When the year began, I predicted that three topics would be in the news: augmented reality, automated assistants, and blockchain. To a certain degree, this was true, but in the end, I wrote about three topics that proved even more popular: Big Data, Privacy, and Security; Genetic Engineering; and Cutting the Cord.

Big Data, Privacy, and Security: It's a sad commentary on the state of the computing industry that almost every day we hear about a new data breach. Facebook, in particular, had a remarkably bad year, with the Cambridge Analytica fiasco, the shambles of testimony before Congress (an experience shared with Google), and the ongoing revelations about the number of third-party developers who had unauthorized access to private user data. I don't see it getting any better in the near future, but I do see movement on the part of the federal government to finally enact data protection legislation along the lines of GDPR in Europe.

Genetic Engineering: The birth of the world's first genetically engineered humans in China marked a turning point in our evolution. The use of the CRISPR toolset for editing our genome is quickly becoming commonplace. The ethical implications of these experiments (which is what they are) are profound, and no one is really sure whether it will lead us to improved health or down the dark road of eugenics.

Cutting the Cord: The number of people who subscribe to traditional cable TV service continues to drop precipitously. There are so many alternatives available now that there's little reason, other than inertia, to stick with the old technology. I was able to watch all of the FIFA World Cup online, and I continue to enjoy programs from around the world via apps installed on my Apple TV. Best of all, there are no advertisements. Believe me, once you've ditched traditional television, you'll never accept suffering through all those annoying interruptions.

***

My Technology Today column is coming to an end. Next week will be the final installment, where I will look back at the notable events that have occurred in the 8-plus years since I started writing the column, way back in October 2010. If you want to keep hearing from me about technology-related matters, send me an email and I'll add you to my mailing list. You can also go online to www.cts.today/join and fill out the simple form. It's free, and you can unsubscribe at any time.

\# \# \#

ALL GOOD THINGS...

So long, and thanks for all the fish

December 28, 2018

As local lad Jim Morrison of "The Doors" once said, "This is the end." This is the final Technology Today column for FLORIDA TODAY. It's not by my choosing, but it's the way it is.

On October 23, 2010, my first headline was "Life with Kindle: So far, so good." I was relating my experience using the (then new) Kindle e-reader from Amazon.com. Today, e-books are an established part of the literary marketplace.

In 2011, I lamented the loss of several computing pioneers, including Steve Jobs. Atlantis completed STS-135 – the final mission of the shuttle program. IBM's Watson supercomputer kicked off modern advancements in artificial intelligence with its victory over humans in Jeopardy!

In 2012, the news was about mobile phones (Apple's iPhone 5 broke sales records – sound familiar?), the emergence of cloud computing as a mainstream platform, and social networks. Facebook went public. Psy's "Gangnam Style" video went viral.

In 2013, I selected privacy, complexity, and fragility as the top three themes of the year. Google Glass was full of promise – until people objected to being spied upon. The digital train wreck known as HealthCare.gov failed to launch. I learned what it was like to lose your iPhone while traveling in Europe.

2014 was the beginning of seemingly daily security vulnerabilities.

Hydraulic fracturing started altering the energy landscape. Doctors Without Borders battled the Ebola outbreak in West Africa with a trial vaccine from Canada.

In 2015, it was the rise of artificial intelligence, the proliferation of Internet-connected devices, and the surging popularity of drones that caught my attention. Siri came first, but Alexa proved to be a more capable digital assistant. Drones shut down Gatwick airport in 2018, illustrating how problematic they have become.

Merriam-Webster's word of the year for 2016 was "surreal." Big Data made its debut. Cybersecurity continued to be a problem: Yahoo! reported over one billion customer email accounts were compromised.

In 2017, you could have made (or lost) a lot of money speculating on cryptocurrencies like Bitcoin. It was also the year when the #MeToo movement started and many powerful people were outed.

This year, Facebook had a remarkably bad year. The first genetically engineered humans were born in China. I continued to enjoy life without cable TV.

I want to thank all my loyal readers. Send me an email and I'll add you to my mailing list. You can also go to www.cts.today/join and fill out the simple form.

I'll close by quoting from *The Hitchhiker's Guide to the Galaxy*: "So long, and thanks for all the fish." It's been a great ride, but it's time to leave before the hyperspace bypass arrives.

# # #

ABOUT THE AUTHOR

Scott Tilley is an emeritus professor at the Florida Institute of Technology, president and founder of the Center for Technology & Society, president and co-founder of Big Data Florida, past president of INCOSE Space Coast, and a Space Coast Writers' Guild Fellow. His recent books include "Systems Analysis & Design" (Cengage, 2020), "SPACE" (Anthology Alliance, 2019), and "Make Technology Great Again" (CTS Press, 2018). He wrote the "Technology Today" column for FLORIDA TODAY from 2010 to 2018. He holds a Ph.D. in computer science from the University of Victoria. Visit his author website at http://www.amazon.com/author/stilley.

www.ingramcontent.com/pod-product-compliance
Lightning Source LLC
LaVergne TN
LVHW052304060326
832902LV00021B/3701